BY THE SAME AUTHOR

The Human Worth of Rigorous Thinking

Mathematical Philosophy

Thinking about Thinking

Mole Philosophy and Other Essays

Science and Religion

The New Infinite and the Old Theology

THE PASTURES OF WONDER

THE REALM OF MATHEMATICS.
AND
THE REALM OF SCIENCE

THE
PASTURES OF WONDER

THE REALM OF MATHEMATICS
AND
THE REALM OF SCIENCE

BY

CASSIUS JACKSON KEYSER
*Adrain Professor Emeritus of Mathematics
Columbia University*

NEW YORK
COLUMBIA UNIVERSITY PRESS
1929

COPYRIGHT 1929
BY COLUMBIA UNIVERSITY PRESS

Published February 1929

Printed in the United States of America

The Torch Press
CEDAR RAPIDS
IOWA

To
the memory of
my wife

PREFACE

THIS essay, written primarily for educated laymen, deals with the two weightiest terms in the vocabulary of rational thought: Mathematics and Science. Its aim is to help the reader to acquire a sound philosophic understanding of what the former term, according to the best critical usage of our time, essentially and distinctively means; and a similar understanding of how the latter term, which has never been given a standard or authoritative definition, ought to be defined.

The propriety of the title will, I trust, become sufficiently evident to anyone who reads the essay attentively. Perhaps no other event in the long development of organic life has been quite so significant and so fruitful of good as the advent of Wonder. "For it is owing to their wonder," said Aristotle, "that men both now begin and at first began to philosophize." Wonder is the great question-asker. Answers are propositions. These are of two basic types — the Categorical and the Hypothetical — and belong to the one or to the other according as the corresponding questions relate to the world of the Actual or to that of the Possible. But for the two great propositional realms — that of categoricals and that of hypotheticals — intellectual curiosity could have no means of subsistence and, were they to fail, would quickly perish by starvation. They are indeed the Pastures of Wonder.

The book is not a "Story" nor a "Romance" nor a 'jazzy' attempt at popularization. Clarity, however, I have tried to achieve, as being doubtless the highest stylistic obligation of an author to a reader whose strongest desire is to understand.

C. J. K.

January, 1929.

CONTENTS

PREFACE ix

INTRODUCTION 1

PART I: THE REALM OF MATHEMATICS . . 21

    The essential nature of mathematics as disclosed by centuries of criticism; hypothetico-deductive character of mathematical thinking; propositions and meaningless statements; implication, deduction, inference.

    The natural classification of propositions as hypothetical and categorical; the two meanings of true and false; definition of mathematical proposition and of mathematics; sheer mathematical thinking not concerned with any specific kind, but applicable to all kinds, of subject-matter; mathematics as form without content; postulates and theorems; impliers and implicates; hypotheses and conclusions; signs and symbols essential to the progress but not to the existence of mathematics; what Euclid would say were he here today; mathematical propositions established deductively, not empirically or pragmatically; the distinction of propositional content and propositional form; having, and only seeming to have, content; implication, or deducibility, a relation pertaining to propositional forms and not to content; inference led astray by content; ineptitude of the terms 'pure' mathematics and 'applied' mathematics.

    Undefined and defined terms; the former essential to discourse, the latter an economic but not a logical necessity; critical importance of distinction between definition and description; the latter the sole means of indicating the presence and kind of subject-matter in discourse not purely formal.

    Dialectic demonstration of the general availability of mathematical thinking.

    Sheer mathematics, essentially and exclusively concerned with the world of the Possible, is *applicable* to the world of the Actual as part of the Possible.

PART II: THE REALM OF SCIENCE . . . 103

A definition of science proposed as a desideratum and submitted to criticism pro and con; the emotive function and the logical function as the two major and all-embracing functions of speech; the ideal of a logically perfect language; some of the necessary conditions for such a language; concerning the logical defects of existing languages; concerning the unattainability of genuine ideals.

The non-existence of a standard or authoritative definition of science; use of the term in a hardly numerable variety of nebulous senses devoid of a common element of meaning; the scandalous situation chargeable in a measure to scientific men; attempts to define science exemplified and characterized; the desideratum stated.

The establishment of categorical propositions (about the actual world) and of hypothetical propositions (about the possible world) viewed as the two all-embracing concerns of the human intellect. Formal definition of science as an enterprise and as a body of established categorical propositions; concerning the meaning of the phrase, an established proposition.

Concerning the advantages and the supposable disadvantages of the proposed definition of science. What *the* scientific method would be according to the proposed definition. Consideration of the fact that, according to the definition, the domain of science would exclude mathematics and include present-day philosophy. Mathematics as an edifice of hypotheticals; science as an edifice of categoricals; either of the two may enter the other as instrument but not as constituent.

Examination and condemnation of the tradition that mathematics is a *model* for science.

Indication of innumerable propositions that, being empirically established categoricals, are genuinely scientific but are erroneously called mathematical.

The need of enlarging the current conception of a scientific laboratory.

Panthetics (from *pan* and *thesis*) proposed as a suitable name to designate that enterprise which embraces both Mathematics and Science.

INDEX 201

INTRODUCTION

INTRODUCTION

IT IS the principal aim of this work to answer two questions:
 (1) What is Mathematics?
 (2) What meaning ought to be given the term Science?

The book is accordingly composed of two parts:
 I. The Realm of Mathematics.
 II. The Realm of Science.

Of these parts the former is a report; the latter, a proposal. The meaning of this statement will appear presently.

For more than 2000 years philosophers and mathematicians endeavored to ascertain what it is that characterizes mathematics and determines its proper place among the other cardinal enterprises of the human spirit. The answer, found only a few years ago, was a great discovery, one of the very greatest in the history of thought. Thus far it has come to the knowledge of hardly any except certain mathematicians and a very few philosophers. Yet it is an answer that can be rendered intelligible to all educated men and women and one that ought to be of interest to them as a very significant contribution to modern culture and general enlightenment. I have said that Part I is a report. It is a report of that answer, of that discovery: a statement and interpretation of it, in terms, moreover, that can be readily understood by anyone both able and willing to give the matter a fair measure of disciplined attention.

Regarding the term science the situation is very different. That term has never been defined. It is true that various aspects of what is commonly called science or scientific work or scientific activity have often been more or less aptly described. But description and definition are radically different things — a critical fact to be again signalized and to be more fully evaluated at a later stage. In what is called the literature of science there are to be found many partial descriptions of what science is supposed to be but nowhere in that literature or elsewhere is the term defined in a way generally recognized in competent circles as adequate or standard or authoritative. If you ask a hundred or a thousand representative men of science to tell you what the term denotes, you will get a hundred or a thousand answers. If you examine the answers you will find that all or nearly all of them are more or less vague and that no two of them are equivalent. And so the question considered in Part II is not what science *is*, as if the meaning of the term had been, like that of mathematics, already fixed and, therefore, required only to be reported and explained. The question is not: What does the term now signify? For, inasmuch as it has never been defined, it has never had, and does not now have, any definite signification. The question is: What *ought* the term to signify? What meaning ought we to give it? The sense of "ought" as here used is doubtless clear enough: the meaning "we ought to give" the term is the meaning which it would be most helpful, most expedient, to give it. Part II is, then, as I have said, not a report, but a proposal. It proposes a *definition* for the term Science, submits it for mature and open-minded consideration, and examines it for its pros and cons with as much candor and circumspection as the writer has been able to command.

A goal is one thing; a journey thereto is another. Neither or both of them may be attractive; or either may be so and the other one not.

As already indicated, the goals to which it is the aim of this book to conduct the reader are two: a thorough understanding of what it is that the term Mathematics, taken strictly in accord with the best critical usage, now signifies; and a thorough understanding of the desirability and feasibility of assigning to the undefined term Science a signification as well based and as definite and clear as that of mathematics or of any other term in speech. It is conceivable that a reader, even if he have little or no interest in either of the goals as such, may yet be sufficiently interested in one or the other or both of the thought-journeys by which the goals are reached. For the journeys may be taken in leisurely fashion, deliberately, hardly otherwise; the air is sufficiently dry and

Introduction

cool, the sky serene; and there is scenery along the way, some of it new perhaps and some of it old but seemingly new because seen in new connections from a new point of view.

Hence it seems worthwhile to sketch briefly here in advance the general direction and the principal features of the course that the essay has taken and to indicate at the same time some of the course-determining considerations.

THE COURSE OF PART I

Notwithstanding the radical difference between mathematical method and the 'method of science,' it is found that mathematics and science cannot be fundamentally discriminated or defined in terms of their respective methods but only in terms of the propositional types with which they are respectively concerned. Consequently the essay begins with a study of propositions, their nature, their basic types, their mutual relations and especially the great relation of Implication, which often serves to bind a host of propositions into a logically organic autonomous whole, system, or doctrine.

Any question whatever leads sooner or later to a proposition purporting to answer it, and every such proposition belongs to one or the other of two mutually exclusive and basic types. If the question be one concerning the make-up of the actual world — the world of Actuality —, the answering proposition will be Categorical, asserting outright that *such-and-such is the case*. But if the question be one concerning the constitution of the infinitely vaster world of Possibility, the answering proposition will be Hypothetical, asserting that, *if such-and-such supposable things were actual, then such-and-such other things would of necessity be so, too*. Thus it becomes evident that, corresponding to the two great propositional types, the knowledge-seeking activity of man presents two grand divisions: quest of categoricals wherewith to describe the actual world; and quest of hypotheticals wherewith to describe the world of possibility.

Very noteworthy is the obvious fact that the matters mentioned are not artificial but are natural. The knowledge-seeking activity of man is natural. For, as Aristotle said the other day, "All men by nature desire to know." Intellectual curiosity, or Wonder, the question-

Introduction

asker, finest of the gifts and distinguishing marks of man, is natural. The realm of propositions, wherein all answers must be sought, is a natural realm. And so, too, are the great types there found, that of the categoricals and that of the hypotheticals. These propositional types are, then, a perfectly natural basis for defining Science and Mathematics — two distinct enterprises that, taken together, are seen to constitute the whole effort of our human kind to understand the World.

The current conceptions of hypothetical and categorical are rejected as superficial and misleading, and the terms are redefined. A proposition of either kind may or may not have the *if-then* form, which, though often convenient, is never essential.

In connection with hypotheticals, which it is the characteristic aim of mathematics to establish, the rôle of implication is specially signalized. It seems impossible to overestimate the importance of that relation. For without implication, we could not infer, deductive reasoning would be impossible, and the highest forms of life would be sub-human. I repeat here what I have said elsewhere: "If one could in some wise suddenly reveal clearly to mankind the full extent to which human culture, human achievements, human civilization in all its aspects and dimensions have depended upon the silent service rendered by the subtle relation which we denote in English by the little word *imply*, the vision would fill its beholders with ineffable wonder and awe; so little do we, ordinarily, sense the immeasurable gravity of a great imponderable principle." It is implication that makes possible the very existence of mathematics, for, without implication, hypothetical propositions would be meaningless — they would not even exist.

The cardinal terms, mathematical proposition and established proposition, are sharply defined. The essential distinction between saying that a proposition is true and saying that it is established is drawn and emphasized. And it is seen that the adjective true (or false) has two essentially different significations according as it is applied to a hypothetical proposition or to a categorical one.

It is shown that mathematical symbols, however indispensable to the progress of mathematics, are, contrary to popular belief, not essential to the existence of mathematics. The methodological and epistemological importance of distinguishing between propositional content and propositional form is accentuated. The same is true regarding the subtle but very real distinction between 'having' content and 'seeming to have' it.

In virtue of such distinctions it turns out that, among significant by-products of the discussion, there is the following two-fold myth-destroying thesis: (1) That in sheer mathematical thinking we are not thinking of any specific kind of subject-matter; and (2) That, just because sheer mathematical thinking is independent of any and every specific kind of subject-matter, it is applicable to *all* kinds. In illustration of that towering thesis, a very simple specimen of sheer mathematical propositions is applied now to one and now to another of various subject-matters — those of biology, ethics, mysticism, theology, and jurisprudence. The contention that there is or can be a kind of subject-matter to which mathematical thinking cannot be applied is dialectically shown to involve a contradiction. It is made evident that the current terms, 'pure' mathematics and 'applied' mathematics are misnomers, inherited from the stone age of mathematics. The continued use of those world-befuddling terms by mathematicians is unworthy of enlightened men.

It goes without saying that, in an essay designed to lay bare the distinctive character of mathematical thinking, the nature and the office of postulates are not overlooked. To correct a very unfortunate error, as old as history and world-wide today, the fact is stressed that in any mathematical work (for instance in Euclid's *Elements*) are found many so-called mathematical propositions that are strictly not mathematical propositions at all. For example, the familiar Euclidean proposition regarding the square on the hypotenuse of a right triangle is not a mathematical proposition. What is (in that connection) mathematical is the proposition that the other one is implied by certain stated postulates.

The radical but commonly unregarded distinction between description and definition is signalized as of immense critical importance; their respective rôles are explained and exemplified. The function of undefined terms in discourse is shown to be a logical necessity while that of defined terms, though not a logical necessity, is a practical or economic one. There emerges a major thesis: Any discourse whatever, whether mathematical or non-mathematical, is, and of necessity must be, discourse about terms or symbols that, however much they are or may be described, remain ultimately undefined.

THE COURSE OF PART II

The reader is invited to consider maturely with open mind a certain proposal — that of assigning to the great term Science a satisfactory signification; he is not invited to consider anything so meagre as a mere attempt to add another item to the existing list of dictionary definitions; he is invited to share in the endeavor to do an important bit of critical work in the light, and under the inspiration and authority, of a great ideal—that of a logically perfect language. Being a genuine ideal, it is, of course, not to be fully realized. For genuine ideals are not of the nature of a goal — something to be reached; they are perfections to be pursued endlessly, limits that are always beyond, to be more and more nearly approached but never attained.

Attention is drawn to the highly significant fact that, in English and the other great languages, the galvanic or emotive function of speech — the power of words to engender and to express feeling — is far more developed than is the logical function — the power of words to articulate, concatenate, and communicate thought. The conception of a logically perfect language is the conception of a language equipped to organize thought soundly, in accord with the laws of thought, and to communicate it unambiguously — a conception compared with which existing languages are found to be almost unbelievably defective. Some of the more evident conditions are given which a language, in order to be logically perfect, would have to satisfy, one of them being that no term in the language could stand for more than one object of thought, have multiple meaning or ambiguous signification.

Obviously the ideal condition just now mentioned as essential is violated, grossly violated, by the term Science, the term being currently employed, not only in an innumerable variety of senses, exhibiting every possible degree of nebulosity, but even in senses having nothing whatever in common, as shown, for example, by the familiar phrases, Christian Science and mathematical science. For the scandalous situation it is found that scientific men are themselves partly to blame, as having been too little concerned to fix the meaning of their great term or to protect it against degrading misappropriations and abuse by the cunning or the vulgar.

It is proposed to define a scientific proposition to be an established categorical proposition; the phrase, established proposition, is, as before said, carefully defined. The proposed definition of science is formulated as follows: *Regarded as an enterprise Science is characterized by its aim, which is that of establishing Categorical propositions.* There is added a virtually equivalent definition of science regarded as a body of achievements.

The merits of the definition are set forth. The chief advantages that would accrue from general adoption of it are indicated and assessed. But are these sufficient to win for it the general approval of 'scientific' men? It is almost certain that the typical 'scientific' expert's *first* impression will be that they are not sufficient. There are, he will say, fatal objections to be urged. He is invited to urge them. And what are they? There are several but they are included in two. One of them is that the definition is too *narrow* because, in violence to a world-wide hoary tradition, it excludes mathematics from the domain of science; the other one is that it is too *broad* because it includes in that domain the speculations of philosophy.

In response to the first objection it is shown:

(1) That, if the natures of mathematics, deduction, and experimentation had been understood in the remote past when that now hoary tradition began, it could not have begun.

(2) That innumerable so-called mathemat-

ical propositions—such, for example, as those constituting the arithmetic of the counting-house and the geometry of the carpenter — are not mathematical but are strictly scientific in even the critic's sense of scientific, for they were discovered and established by observation and experiment long before mathematicians succeeded (only recently) in deducing them from postulates.

(3) That no mathematical proposition states a physical fact or biological fact or ethical fact or any fact peculiar to any specific sort of subject-matter; that, on the contrary, a mathematical proposition merely asserts an implication between two propositional forms, regardless of content.

(4) That a mathematical proposition, though employed in, say, a physical or a biological research, is not a physics or a biology proposition and cannot be authentically asserted as such but only as a tool employed.

(5) That just as a mathematical treatise may contain, but cannot *assert*, categorical propositions, however well established in science; so a scientific treatise may contain,

but cannot *assert*, hypothetical propositions, however well estblished in mathematics.

Part II (and therewith the book) closes with a statement of, and a response to, the 'scientific' expert's reasons for spurning a definition of science that will require him to regard philosophy seriously as a genuine branch of science. He charges that philosophy (a) asserts but does not establish propositions; (b) depends too exclusively on ratiocination, neglecting facts; (c) is unintelligible or meaningless because of inadequate definition of her major terms; and (d), worst of all, does not employ the laboratory method of observation, experimentation, hypothesis and verification.

In response it is shown:

(1) That the propositions which it is the aim of present-day philosophy to establish are, unlike those of mathematics but like those of physics or other 'scientific' branch, categorical propositions.

(2) That innumerable propositions among those that must be credited to philosophy have had, many of them for long periods of time, the status of established propositions and that many of them have that status today.

(3) That such philosophers as have relied too exclusively upon ratiocination are numerically matched in 'scientific' circles by mere fact-gatherers neglecting interpretation.

(4) That important terms of indeterminate signification are neither more numerous nor more conspicuous nor more baffling in the literature of philosophy than in that of 'science.'

(5) That, rightly understood, a laboratory exists wherever *thinking* occurs — wherever a human being observes, identifies, remembers, imagines, conceives, discriminates, compares, analyzes, combines, reasons and judges; and when the thinking aims, consciously or unconsciously, to establish some categorical proposition respecting no matter what subject-matter or aspect of the world, the laboratory ought to be regarded a scientific one.

In closing this introduction it seems fair to say that for grasping the book's main thesis in all its bearings and for feeling the full force of what is said in defense of it there is demanded a certain magnanimity. Sensitivity to minute points, nice distinctions and delicate considerations, though it is essential, is far from sufficient; there must be also a sense for what is large, sweeping, momentous; and especially must there be clear discernment of the amplitude and fundamentality of the background on which the entire discussion is projected.

PART I
THE REALM OF MATHEMATICS

THE REALM OF MATHEMATICS

WHAT is Mathematics?

It required 500,000 or more years of preparation before the race of Man was sufficiently advanced in thought to ask that question. After the question was at length propounded mathematicians and philosophers required twenty-five centuries to answer it. For, though serious endeavor to find the answer began even before the days of Plato, yet the discovery of what mathematics essentially and distinctively is is a very recent achievement. The story of that endeavor through the intervening years would make a deep and edifying as well as romantic chapter in a Critical History of Thought. I am not about to recount the tale. It is not with the story of the long research but with the outcome of it that I purpose to deal.

Now that the answer to the great question has at length been found it is a genuine pleasure to be able to assure the reader at the outset that the answer is not one of those which none but expert or professional mathematicians can understand. On the contrary it can be made sufficiently intelligible to any man or woman who is able and willing to give it a fair measure of disciplined attention.

The answer in question has been rendered in various substantially equivalent forms. Of these one of the best, especially in point of neatness and suggestiveness, is a formulation given a few years ago by a distinguished Italian mathematician, Mario Pieri. Pieri's answer is: Mathematics is the hypothetico-deductive science. Had he been asked to tell what mathematical thinking is, he would doubtless have said: Mathematical thinking is hypothetico-deductive thinking.

HALF THE SECRET OF PHILOSOPHY

What does the answer mean? Let no one be dismayed by the frightful aspect of that hyphenated adjective—hypothetico-deductive. Its significance may not be quite so clear as the serene blue of an Italian sky but a little reflection will suffice to make it so. Half the secret of philosophy, said Leibniz, is to treat the familiar as unfamiliar. There is no wiser maxim and none fitter for present use. In studying the sense of Pieri's words, in trying to understand his fine characterization of mathematics and mathematical thinking, it will be very helpful to look first, a bit keenly if we can, at a few old and familiar items as if they were new and strange.

PROPOSITIONS, SENSE AND NONSENSE

One of the items is what is called a proposition. Propositions occur in much of our silent meditation, in much of what we say, in much of what we hear, in much of what we read. Not many important terms are so familiar as the term proposition. How many people, do you suppose, have ever really wondered what a proposition is? No doubt the number of them is very small, for it is not easy to *think* about any matter that is very familiar. Intellectual curiosity is seldom piqued save by the odd, the unexpected, the novel or the strange; which is a way of saying that, though people are many, philosophers are few.

What *is* a proposition? I know of no better answer than this: A proposition is a statement having one or the other of two qualities,

26   *The Realm of Mathematics*

that of being true or that of being false. Some statements are meaningless, or nonsensical; they are neither true nor false and so are not propositions. Here is a fair specimen: "Even God is God and not God as well." Have *you* never uttered so meaningless a statement? You have unwittingly and solemnly uttered thousands that are just as bad or worse and so have I. Nonsensical statements, erroneously taken for propositions, abound in human discourse, both in that of the wise and in that of the foolish. On this account a very great deal of our human discourse is itself nonsensical; much of it, though high in its aim, confident, sincere, solemn, stately, seemingly well reasoned, read or heard with admiration and awe, gravely cited as authority, is nevertheless no expression of thought but is mere chattering.[1] The matter here barely touched upon in passing, I mean the distinction of 'true or false' and 'nonsensical,' is fundamental and will be discussed more fully at a later stage.

IMPLICATION, DEDUCTION, INFERENCE

Another of the familiar items which, as I

[1] In this connection see pages 38-44 of Keyser's *Mole Philosophy* (Dutton & Co.).

have said, it will be advantageous to look upon as strange is the fact that propositions are so related among themselves that some propositions *imply* other ones. A most fortunate circumstance! For, were it not for the fact of such implication, there could be no reasoning by inference, no logically organized science, no mathematics, no deductive logic, no human civilization, no humankind. Life might be possible in such a world but not human life. So vital a relation as that denoted by the little word, implies, essential to the very life of a rational organism, is worthy of some attention. Let us give it a little. What I am about to say is not very hard to understand but it does require to be pondered a bit.

Suppose p and q are taken to denote two given propositions. Then the statement, p implies q, is a third proposition. What is it that one asserts in asserting such a proposition? I am going to assume that the answer is this: *To assert that p implies q is equivalent to asserting that p and q are formally so related that q can be logically deduced from p.* What the answer means may, I believe, be made sufficiently clear by a familiar geometric example. For it is highly probable that the reader has had, in high school or in college, some instruction in geometry, at least a little of it. Such a reader will remember that certain propositions (which were probably called theorems) were deduced from certain initial propositions (which may have been called axioms). And now the example is this: if we let p denote the latter propositions and q one of the former, then to say that p implies q is just a way of saying that q is deducible from p.

TWO BASIC TERMS NEWLY DEFINED AND WHY

It is common and often convenient, as the reader knows, to state a proposition of the form, p implies q, in another form, which we may call the If-Then form: if p, then q. This form of statement leads naturally to the consideration of two terms which, when suitably defined, are of the gravest importance. I mean the terms: *hypothetical* proposition and *categorical* proposition. Because of their exceeding importance it is necessary to make very clear what it is that they are respectively to signify as used in the following pages. I have said "as used in the following pages" because, for reasons to be presently stated, I intend to reject the usual definition of the terms and to replace it by a radically different one.

How are the terms in question usually defined? In most of the books dealing with the logical theory of propositions the two terms are virtually defined as follows: A proposition is said to be hypothetical if it has the if-then form and categorical if it has not. As thus defined the terms are, for scientific use, worse than worthless, for a little reflection will show that the definition diverts attention from what is essential by stressing what is not and that it is thus not only shallow but misleading. To see that it is so consider propositions of the form — if p, then q — where, as before, p and q are themselves propositions. When asserting such an if-then proposition, are you intending to assert thereby that q is logically, or formally, deducible from p? A glance at a few particular examples will suffice to show that sometimes you are and sometimes you are not. By recalling the geometric example cited above it becomes evident that, in that instance, when you assert the proposition — if p, then q — you are intending to assert that p implies q, that is, you are intending to assert that q is deducible from p. For other examples of if-then propositions, take the following, chosen almost at random:

(1) If Socrates was an Athenian and all Athenians were Christians, then Socrates was a Christian.

(2) If it lightens, then it will thunder.

(3) If a man dies, then he will live in another world.

(4) If Smith libels Jones, then Jones will sue for damage.

(5) If A gives B a horse, then B will give A $100.

(6) If you pray for me, then I will pray for you.

(7) If I live in Albany and Albany is in New York, then I live in New York.

Note that each of the propositions is of the form — if p, then q — p denoting in each example what follows 'if' and precedes the comma, and q denoting what follows 'then.' It is perfectly evident that, when you assert (1) or

(7), you are to be understood as intending to assert thereby, as in case of the above-given geometric example, that q is logically deducible[2] from p. It is equally evident that, when you assert (2) or (3) or (4) or (5) or (6), you are not to be understood as inteding to assert any such logical deducibility of q from p. Thunder is often, perhaps always, a *physical* consequence of lightning but fancy yourself solemnly asserting that it is a logical consequence, a syllogistic consequence, of lightning. Think of intending to assert seriously that the proposition, I will pray for you, can be logically deduced from the proposition, you pray for me; or that the proposition, B will give A $100, is logically deducible from the proposition, A gives B a horse.

Let us now take stock a little. We have seen that to assert a proposition of the form, p im-

[2] Of course I mean by help of such logical principles as: (a) A cannot be both B and not-B; (b) An hypothesis having a false consequence is false; (c) The negative of a true or false proposition is false or true. For such principles or their equivalents are essential to all logical deduction. Hence they are usually not stated but are employed tacitly. When we say that q is deducible from p, we mean deducible from p joined with such principles. In other words, when we say that p implies q we mean that p and those principles conjointly imply q.

plies q, is equivalent to asserting that q is logically deducible from p. We have also seen that an if-then proposition — as (1), for example, or (7) — may be such that to assert it is likewise equivalent to asserting that q is deducible from p. In such a case it is natural and legitimate to say, and I am going to say, that the proposition, p implies q, and the proposition — if p, then q — are equivalent. Now, according to the usual definition, above cited, of the terms, hypothetical proposition and categorical proposition, the foregoing proposition, p implies q, not being in the if-then form, is categorical while its equivalent — if p, then q — is hypothetical. And so, according to that definition, two propositions, though they be *equivalent*, may be one of them hypothetical and the other one categorical. That is one reason why I say that the definition in question is shallow, so shallow as to be scientifically and philosophically worthless or worse.

TWO RADICALLY DIFFERENT MEANINGS OF TRUE OR FALSE

But there is another reason, a yet profounder reason, for discarding that definition. It is a reason connected with the significance of the words, true and false. The reader may be aware of the fact that each of these adjectives has two radically different meanings according as it is applied to propositions of one type or to those of another. Let us recall these two meanings. If you say that a certain proposition, p implies q, is true, what do you mean by "true"? You mean that q is logically deducible from p, and you mean nothing else. Similarly if you say that the proposition is false, you mean by "false" that q cannot be deduced from p. We may conveniently call this sense of the word, true, or the word, false, the *Dsense* (D suggesting deducible). Let us now suppose that some propsition — if p, then q — is, like (1) or (7), equivalent to the proposition, p implies q. It is evident that the word, true, or the word, false, is applicable to such an if-then proposition in the D-sense and in no other. On the other hand, consider, for example, the proposition (2) — if it lightens, then it will thunder. You may assert that it is true or assert that it is false. In neither case will you be using the adjective in its D-sense for it is obvious that the evidence for your assertion is not of the *deductive* kind but is, ultimately, observational or *empirical*. We may conveniently call this sense of the word, true, or the word, false, the *E-sense* (E suggesting empirical). It is clear that, if I apply either of the two adjectives to such a proposition as (8)—the orbit of the Earth is an ellipse — or to its if-then equivalent (9) — if O denotes the Earth's orbit, then O denotes an ellipse — I shall thus be employing the adjective, not in its D-sense, but in its E-sense.

The reader is now in a position to see that the current definition by which a proposition is hypothetical or categorical according as it has or has not the if-then form does not discriminate propositions with any reference to the fundamentally important matter of the sense in which they may be significantly said to be true or false. For, in the light of the foregoing discussion, it is evident that, by the definition in question, a proposition that is true or false in the D-sense of the adjective may be hypothetical, like (1) or (7) for example, or it may be categorical, like the proposition, p implies q; a proposition, true or false in the E-sense, may also be either hypothetical or categorical, the fact being shown by (2) and (8) for, by the definition, (2) is hypothetical and (8) is categorical; moreover, two equivalent propositions of which, by the definition, one is hypothetical and the other categorical, may be true or false in the E-sense, like (9) and (8), for example, or in the D-sense, like (1) and the proposition (10) — 'Socrates was an Athenian and all Athenians were Christians' implies 'Socrates was a Christian.'

The reasons given are sufficient, I believe, to convince any one that the definition under discussion is so shallow, so lacking in penetration, that the terms, hypothetical proposition and categorical proposition, have, when thus defined, but little, if any, importance. It is my intention, as I have said, to discard that current definition and to replace it by one which, I believe, will endow the two terms in question with critical significance unsurpassed by any terms in scientific methodology or even the theory of knowledge.

The new definition is as follows:

If a proposition, P, is such that to assert it is equivalent to asserting that a proposition q is logically deducible from a proposition p or — what is tantamount — that p implies q, then P is a hypothetical proposition; in the contrary case, P is a categorical proposition.

Such is the sense in which the two terms are to be understood in this book. If P be hypothetical, p is called the *hypothesis* or the *implier* and q is called the *conclusion* or the *implicate*. Note that henceforth the two kinds of proposition are not to be discriminated by the presence or absence of the if-then form, for, by the new definition, a hypothetical proposition may have that form, as shown by (1) or (7), or it may not have it, as shown by (10) or by any proposition of the form, p implies q. Similarly a categorical proposition may have the if-then form, as shown by (2) for example, or it may not, as shown by (8) for example. Note, too, that, by the new definition, if two propositions be equivalent, either they are

The Realm of Mathematics

both of them hypothetical, a fact exemplified by (1) and (10), or they are both of them categorical, a fact exemplified by (8) and (9). Moreover, the following four facts are to be very specially noted: a proposition is hypothetical if it be true or false in the D-sense of the adjective; conversely, if a proposition be true or false in that sense, it is hypothetical; a proposition is categorical if it be true or false in the E-sense; and, conversely, if a proposition be true or false in that sense, it is categorical.

The last four propositions make it evident that our new definition is precisely equivalent to the following one:

A proposition is hypothetical or categorical according as it is true (or false) in the D-sense or in the E-sense.

Inasmuch as the two new definitions are equivalent it is logically indifferent, solely a matter of covenience or taste, which of them we employ.

In view of what has been said it is hardly necessary to caution the reader against supposing that the hypothesis or implier p, or that the conclusion or implicate q, of a hypothetical proposition, must be a single proposition. As a matter of fact p may consist, and usually does consist, of several propositions — a small set or system of them, as we say; and q may be but one proposition or a thousand of them. We are to understand that, if p and q each denotes a system of one or more propositions, the statement, p implies q, or any equivalent statement, is just a short way of saying that the propositions of the former system conjointly imply those of the latter system. Such is, of course, the case in the above-cited geometric example where p is a set of axioms and q is one or more theorems deduced from them.

If a hypothetical proposition be true, the fact may be obvious, as in the foregoing examples (1) and (7), or it may require much ingenuity or even great genius to show it. The process of showing it, as the reader knows, is called the proof or the demonstration of the proposition. He knows, too, and should pause to realize vividly, that the demonstration of a hypothetical proposition, p implies q, always consists in *deducing* q from the hypothesis, *p*. The process is one of deductive logic. With the process of such deduction the reader has some familiarity, acquired by experience in following or in forging algebraic or geometric demonstrations in high school or college or elsewhere. I shall deal further with the matter as occasion may require.

WHAT A MATHEMATICAL PROPOSITION IS

We are now prepared to answer the very important question: What is a mathematical proposition? The answer is: A mathematical proposition is a hypothetical proposition that is regarded by the mathematical world as having been demonstrated. In other words, it is a hypothetical proposition whose conclusion or implicate, q, is regarded by the competent as having been logically deduced from the proposition's hypothesis, or implier, *p*. "Mathematics," said Pieri, "is the hypothetico-deductive science." I venture to believe that we are now beginning to see what he meant. I have said "beginning to see," for the *full* significance of his *mot* is too profound, too subtle and too vast to be so quickly disclosed, and, for an adequate understanding of it, our meditation has a fairly long course yet to run.

TWO MYTH-DESTROYING FACTS AND THEIR SIGNIFICANCE

The preceding paragraph contains two definitions of major importance: a definition of the term, Mathematical Proposition, and a definition of Mathematics. There are two facts about them which we must not fail to observe, for the facts in question are fatal, or ought to be fatal, to a pair of ages-old and still reigning myths regarding the essential nature and the scope of mathematical method. One of the facts is that neither the definition of mathematical proposition nor that of mathematics says anything about quantities or about numbers or about geometric entities or about any other specific kind of subject-matter. The other fact is that neither of the two definitions says anything about those strange, repellent, world-frightening signs and symbols which increasinly abound in mathematical literature and which are, commonly, about the only things of which the word mathematics recalls even so much as a vague and jumbled impression. The critical significance of the two facts is fundamental. Let us examine it somewhat attentively.

SHEER MATHEMATICS IS FORM WITHOUT CONTENT

The first one of the mentioned facts signifies that, when thinking mathematically, we need not be thinking about quantities or magnitudes or about numbers or about geometric entities or spacial configurations or about any other specific kind of subject-matter; what is much more it signifies that, when thinking mathematically, we are thinking in a way which, because it is independent of what is peculiar to any kind of subject-matter, is *applicable* to *all* kinds — available, that is, in every field of thought. The thesis just stated regarding the general availability of mathematical thinking is so important for the prosperous conduct of human life that its importance cannot be exaggerated. It is among the major theses of this book and will come to light again and again in the following pages. For the present I will merely exemplify it by a simple example familiar to all.

Consider the hypothetical proposition: If John Doe was in Chicago at midnight of June 30, 1926, then he did not at that time stab Richard Roe in New York City. Ordinarily the proposition would be regarded as obviously true. Yet, strictly taken, it is not true, for the conclusion cannot be deduced from the stated hypothesis. The deduction becomes possible if and only if the stated hypothesis be enlarged by adding to it certain propositions which the defendant's counsel might think it unnecessary to state explicitly because a juror would unconsciously take them for granted. I mean such propositions as that the alleged stabbing required the presence of the stabber at the time and place of the deed and that the two cities mentioned are such that Doe could not have been in both of them at the same time, which of course he might have been were the cities overlapping. If the stated hypothesis be thus rightly enlarged, the deduction in question becomes possible and the proposition true and logically demonstrable. It is thus evident that every *alibi* defense involves the application of a genuine bit of mathematics, a genuine bit of hypothetico-deductive thinking. By a little observation and reflection readers can discover for themselves that many similar examples occur, in more or less disguised and often imperfect form, here, there and yonder, in all connections and situations, high or low, near or remote, where human beings have tried to *infer*. For every attempt to infer a proposition from other propositions is an attempt to think mathematically and whenever, as frequently happens, an attempt to infer is successful, some mathematical thinking has been done. The fact is that the impulse to such thinking is a normal impulse, so nearly universal that even a 'fundamentalist' occasionally feels it in some measure as when, for example, he says: "If the biological account of the descent of man is true, then the biblical story of the creation of man is false." *Mythical* he should say instead of "false," for, as we have seen, nonsensical statements, not being propositions, are neither true nor false. Is the distinction perhaps too fundamental for 'fundamentalists'?

THE ESSENCE OF MATHEMATICS IS NOT IN ITS SYMBOLS

I have already drawn attention to the fact that the definition of a mathematical proposition and the definition of mathematics are both of them silent respecting those peculiar signs and symbols which professional mathematicians so much employ and without which, it is commonly believed, mathematical thinking would be impossible. What does that silence signify? It signifies that the mentioned belief is a myth. Mathematical signs and symbols are nothing but linguistic devices gradually invented for the purpose of economising intellectual energy and, because they serve that purpose so well, their use is highly expedient. But a vast deal of mathematical thinking was done before they were invented and much of it is now done without their use, by means of the words or symbols of ordinary speech, just as agriculture existed for ages before the invention of modern agricultural machinery and is even now extensively carried on without the use of such machinery, by means of primitive implements. For mathematical purposes ordinary words are primitive instruments. The economic power of mathematical symbols is indeed very great, so great that mathematicians have been thereby enabled to construct many a doctrine that they could not have constructed without using them. And though such a doctrine, once it has been thus constructed, could by great labor be translated into ordinary language, yet the resulting discourse would be so prolix, involved, and cumbrous that none but a god could read it understandingly and no god would do it unless he were a divine fool. Notwithstanding the immense service rendered by the symbols in question, it is no more true to say that without them there could be no mathematical thinking than to say that without the modern means of passenger transportation there could be no travelling or that fighting would be impossible were there no modern instruments of war.

The message which it is the aim of this book to convey must remain in large measure unintelligible to anyone who fails to grasp and apply the respective meanings of certain cardinal terms. One of these is the term, mathematical proposition. I have defined it carefully and deliberately. The definition is not familiar. Moreover it does not accord with the sense or senses, vague or contradictory or both, in which the term is employed in common parlance and, frequently, even in the speech of mathematicians themselves. It would be strange if the reader were not at first somewhat bewildered. I wish to guard him against such bewilderment, to clear it away if it exists, and at the same time to bring the essential nature of mathematics and that of mathematical thinking into clearer light. The definition in question consists of two parts. The bewildering part is the statement that a mathematical proposition is always hypothetical, never categorical. Puzzled by it because it contradicts what he has hitherto been led to believe, a thoughtful reader may desire to challenge the statement.

CHALLENGE BY A THOUGHTFUL READER

"I am not," such a reader may say, "a pro-

The Realm of Mathematics

fessional mathematician but I have had some mathematical instruction and I occasionally turn the pages of a mathematical magazine or book. I know that in any such work are to be found numerous statements which the author has formally set down as propositions, which he submits as having been 'proved,' and which he regards, and expects his readers to regard, as mathematical propositions. Yet many of them are not hypothetical but are categorical. I may as well be specific. I have here a copy of the famous *Elements* of Euclid, of which every one has heard. Opening it, I find, for example, the following proposition, long familiar to the educated world: *The square on the hypotenuse of any right-angled triangle is equal to the sum of the squares on the other two sides.* Certainly that is a mathematical proposition but it is not hypothetical, it is catgorical."

THE READER'S CHALLENGE ANSWERED

The answer to that challenge is as follows:

"The proposition you have cited is, as you have said, categorical but it is not a mathematical proposition; it is only a *part* of one; it is, as we are soon to see, merely the q of a hypothetical proposition — p implies q — whose hypothesis, p, you did not quote. To see that such is the case you have merely to observe what it is that Euclid really did. And what he did is this: in the beginning of his treatise he laid down a small number of propositions, calling some of them postulates and some of them axioms (instead of calling *all* of them postulates or axioms or assumptions as we commonly do today); these initial propositions he neither proved nor pretended to prove; he employed them merely as hypotheses, or impliers; the proposition you have quoted is simply one of their implicates. If we denote it by q and denote by p the postulates from which Euclid deduced it, then the hypotheical proposition, p implies q, or its equivalent — if p, then q — is genuinely mathematical, but q is not and p is not.

WERE EUCLID HERE

"Were Euclid here and were you to ask him the following questions, he would give you the following answers:

" 'Have you proved the propositions p?'

" 'No, I have assumed them, taken them for granted, used them merely as impliers, as hypotheses, as logical *ifs*.'

" 'In your famous book you said that you proved q. Did you mean that you had proved q to be *true*?'

" 'No.'

" 'What did you mean?'

" 'I meant that I had deduced q from p; in other words, I meant that I had shown q to be an implicate of p; in still better words, I meant that I had proved the proposition, p implies q.'

" 'Would it not have been better to say just that explicitly instead of saying what you did say?'

" 'Yes, far better, for then I should not have misled so many, many thousands of innocent people during the centuries since I passed from the bright light of Egypt to the cheerless realm of shades.'

" 'Do you assert that your postulates p are true propositions?'

" 'No.'

" 'Do you assert that q is true?'

" 'I do not.'

" 'Do you assert that the proposition, p implies q, is true?'

" 'I do.'

" 'Which of the propositions found in your *Elements* are mathematical propositions?'

" 'Those and only those having the form, p implies q, or the form — if p, then q — where p denotes the initial [3] postulates and q denotes any proposition deduced therefrom.'

" 'Are those mathematical propositions true?'

" 'Yes.'

" 'Do you say that because they *work* in the pragmatic sense of this term?'

" 'No.'

" 'Why do you say it?'

" 'Because I have demonstrated them, not empirically or pragmatically, for that cannot be done, but logically, deductively.' "

NON-MATHEMATICAL PROPOSITIONS IN MATHEMATICAL WORKS

In and of itself the *Elements* of Euclid is a very great achievement, one of the greatest of

---

[3] Euclid's work is not flawless. He unconsciously used some

the human intellect. But, viewed as a constituent of the immense body of now existing mathematics, it is only a small, though precious, fragment. In the foregoing discussion I have used it in preference to other works merely because of its familiarity and fame. The reader is to understand that what I have said of its method and make-up is essentially true of all mathematical works whatever. The propositions found in any such work fall into two general classes, those which are mathematical and those which are not. The latter class is composed of the postulates (or axioms or assumptions) and the propositions (or theorems as they are often called) that are deduced from them. If T denote one of the theorems and if P denote the postulates from which it was deduced, then the proposition — if P, then T — or P implies T, is mathematical, and all the mathematical propositions of the work are of that kind. When it is said, as in common parlance, that T has been 'proved,' what is meant is that T has been deduced from

postulates that he failed to state. But that fact does not invalidate this reply to the reader's foregoing challenge.

P, not that T has been shown to be true. If T has been thus deduced or is thus deducible, the proposition, P implies T, is true, no matter whether T itself is true or is false.

The foregoing considerations have had it for their aim to bring into clear light what may be called the hypothetico-deductive aspect of mathematics, of mathematical method, of mathematical thinking. That aspect is peculiar to such thinking, essential to it, characteristic of it, but it is not quite sole or exclusive; closely connected with it there is another aspect, which we have yet to examine. Before examining it, we ought, I believe, to pause long enough to hear and consider an objection or question which an attentive reader may, very naturally, feel impelled to raise.

A CRITICAL READER DOUBTS BOTH THE SHEER FORMALITY AND THE UNIVERSAL APPLICABILITY OF MATHEMATICS

Such a reader may say:

"I have grasped firmly, I believe, what is meant by the term, mathematical proposition. I now understand fairly well the significance of Pieri's formula and think it is an admirable, penetrating, revealing *mot*. I begin to appreciate the supreme importance of Implication as the relation which so binds certain propositions to others that the latter can be inferred, or deduced, from the former. I see clearly that without it there could be no logically concatenated discourse, no such thing as a logically organized body of thought, no logically coherent doctrine in any field, no proper life of reason. And I have more than glimpsed the fact that mathematical thinking has for its distinctive function to trace and disclose that curious binding-thread — *implication* — in all its subtly winding ramifications throughout the realm of propositions. All that I freely and gladly grant. But you have said two things that puzzle me. You have said that, when thinking mathematically, we are thinking in a way that is independent of what is peculiar to any kind of subject-matter (such as numbers, for example, or geometric entities); and you have said that such thinking, being thus independent of any specific kind of subject-matter, is *applicable* to *all* kinds. I can see that, if the first statement be true, the second may be true; for, if the first be true, then the thinking found in algebra, say, or in geometry is not sheer mathematical thinking but is an *application* of such thinking to a certain kind of matter — to the properties and relations of numbers, in the one case, and, in the other, to the properties and relations of spacial entities; and thereby one would be prepared to say that, since it is applicable to *some* kinds of subject-matter, it may be that it is applicable to *all* kinds. But I do not see that that first statement is true. On the contrary it seems to me that one cannot think at all, whether mathematically or otherwise, without thinking about some definite sort of subject-matter."

DISTINCTION OF CONTENT AND FORM IS ESSENTIAL

To readers offering that very thoughtful objection I submit the following reply:

"Your criticism imposes upon me a double task. I have to show more clearly than I have done that mathematical thinking does not essentially involve what is peculiar to any kind of subject-matter and that such thinking is applicable to all kinds. To do so it will be sufficient to distinguish between the content and the form of a proposition and then to show that the validity of an inference, or a deduction, depends not at all upon the content of the propositions involved but solely upon their forms.

"To distinguish between content and form, consider the propositions:[4]

(1) Aristotle was a pupil of Plato,
(2) Roosevelt was a rival of Wilson,
(3) X was a Y of Z,
(4) Blood is thicker than water,
(5) Lead is heavier than wool,
(6) X is greater than Y.

The propositions (1), (2), (4) and (5) have content, that of (1) being due to the terms Aristotle, pupil and Plato; that of (2) to the terms Roosevelt, rival and Wilson; that of (4) to the terms blood, thicker and water;

[4] Strictly speaking, such statements as (3) and (6) are not propositions because, owing to the fact that no meanings have been assigned to the symbols X, Y, and Z, the statements are neither true nor false. They are sometimes called *propositional functions*, a term due to Mr. Bertrand Russell and used by me in my *Mathematical Philosophy*. In the present discussion I

that of (5) to the terms lead, heavier and wool. On the other hand, (3) and (6) have no content because X, Y and Z, having been given no meanings, refer to no subject-matter. But *all* of the propositions have form. What is meant? It is a fine insight of Ludwig Wittgenstein[5] that the form of a proposition cannot be expressed but can be shown, not said but seen. We may say that a proposition having no content is itself just form. If it have content, the proposition's form is what remains unchanged in it if we replace its content-giving terms by other such terms (as in passing thus from (1) to (2) or from (4) to (5) or *vice versa*) or by symbols referring to no subject-matter (as in passing from (1) or (2) to (3) and from (4) or (5) to (6). It is obvious that (1) and (2) have the same form as (3) has or is and that (4) and (5) have the same form as (6) has or is but that the form of (1) or (2) or (3) is different from that of (4) or (5) or (6). Of course there are many other

shall call such statements propositions in accord with general usage. The reader will not be misled thereby.

[5] *Tractatus Logico-Philosophicus.*

propositional forms — how many no one knows. A fit subject for research.

DEDUCTION DEPENDS ON FORM ONLY

"I regret that what I have just been saying is so arid. Apart from the fact that, if you had not been considering it, you might have been doing something less worth while, my sole excuse for inviting you to consider it is that your criticism has compelled me to do so, for I know of no better way to make clear the fundamental distinction between propositional content and propositional form. I have now to show that the possibility, process and validity of a deduction, or inference, depend solely upon form, never upon content (or subject-matter). Again the discussion, though not very long, must needs seem to be a little arid unless a reader is genuinely interested in the anatomy of logical thought. There is no use preaching the importance of thinking without telling *how* to think.

"Consider the propositions

(1) X is a Y,

(2) All Y's are Z's,

(3) X is a Z;
(a) A cannot be both B and not-B,
(b) An hypothesis having a false consequence is false,
(c) The negative of a true or a false proposition is false or true.

As you know, it is commonly said that (3) is implied by (1) and (2); in other words, that (3) can be deduced, or inferred, from (1) and (2); but, strictly, it is not so; the deduction, or inference, requires (a), (b) and (c), or their equivalents, in addition to (1) and (2), as we are soon to see; (a), (b) and (c), or their equivalents, being essential to, and consciously or unconsciously used in, every logical demonstration, deduction, inference. Let us actually make the deduction. We may do it as follows.

"If X is not a Z, then by (1) there is a Y that is not a Z; so, by (2), that Y both is and is not a Z; which, by (a) is false; hence, by (b), the hypothesis, that X is not a Z, is false; hence, by (c), its negative, X is a Z, is true.

"We now have a mathematical proposition — denote it by (MP) — which I will state in two ways:

(MP) The propositions (1), (2), (a), (b) and (c) conjointly imply the proposition (3); or

(MP) If (1), (2), (a), (b) and (c), then (3).

"If such discourse bores you, it is not because you are a lover of the humanities. For *no* curiosity can be more humane than that which seeks to understand how humans think when they think mathematically. I now invite you to observe that (1), (2), (3), (a), (b) and (c) are propositional forms and that they are devoid of content, involving, that is, neither quantity nor number nor space nor time nor ethics nor religion nor medicine nor commerce nor politics nor any other specific sort of subject, or subject-matter. If you do, you will see that the above mathematical proposition, (MP), merely states that certain propositional forms imply another such form; and you will see that the thinking essentially involved in establishing (MP) — the thinking by which the required deduction is made — is concerned, not at all with content, but exclusively with form.[6] The fact is that all

[6] One might say that propositional forms are themselves a

mathematical thinking, in so far as it consists in making deductions or, what is tantamount, in justifying inferences, deals with propositional forms and nothing else, but that fact is, oftener than not, disguised, and the disguise is very deceptive. I am trying to strip it off or at all events to render it transparent.

"That the fact *is* disguised, harmfully disguised, and *how*, a familiar example will make clear. Consider the propositions

(4) Socrates is a man,
(5) All men are mortals,
(6) Socrates is a mortal.

Note that these have content as well as form and that their form is the same as that of (1), (2), and (3). Obviously, (6) can be deduced from (4), (5), (a), (b), and (c). Does the deduction depend on the content of (4) and (5)? To see that it does not, write the deduction out and compare it with the above deduction of (3). You will thus see that in the deduction of (6) you have employed 'Socrates,' 'man,' and 'mortal,' not as content-givers, but merely as empty marks like X, Y, and Z;

kind of subject-matter and hence that mathematical thinking always deals with subject-matter. But that is just quibbling.

in other words, you will see that, in making the deduction, you took no account of such facts as that Socrates was an Athenian philosopher, friend of Plato, husband of Xantippe, 'gad-fly,' 'mid-wife,' and so on, or that man is a loving, hating, featherless biped or that a mortal is a living thing doomed to die; and so it is evident that the deduction in question deals, not with the content of (4) and (5) but solely with their forms; since these are the same as the forms of (1) and (2), it is evident that the deduction of (6) and that of (3) are, as deductions, identical, though (6) and (3) are not identical.

CONTENT OFTEN LEADS INFERENCE ASTRAY

"What I am here insisting upon is no trivial matter. Far from it. For, in endeavoring to make deductions from propositions having content, the presence of content-giving terms tends to divert one's thinking, from the forms in which the content is expressed, to the content itself; diversion of attention from form to content always tends to hinder the process of deduction instead of helping it; and that fact goes far in explaining why it is that unsuccessful attempts at deduction, or inference, so abound, not only in our daily cogitations, but in all branches of the literature of thought; as when, for example, from the propositions, a Cleric said we build 'a house of prayer for all people' and the Cleric was a Bishop, we try to infer that the Cleric spoke the truth; or when, from the propositions, Isaac Newton was a great man of science and Newton was a Christian, we try to infer that science and religion do not conflict; or when, from the propositions, the Eighteenth Amendment is a token of national righteousness and Jesus is the Son of God, we try to infer that the wine made by the miracle of Cana was unfermented grape juice; and so on and on *ad infinitum* and *ad nauseam*.

A MATHEMATICAL PROPOSITION IN BIOLOGY

"Now observe that, having deduced (6) as above indicated, you have a mathematical proposition, which may be stated thus: (4), (5), (a), (b), and (c), imply (6); or thus: if (4), (5), (a), (b), and (c), then (6). Observe that this proposition relates to a kind of subject-matter, the *biological* kind; that the proposition results from *applying* to that kind of matter precisely the same thinking as that which established our (MP), which relates to no kind; and that indeed the biological proposition can be obtained from (MP) by merely replacing its meaningless marks, X, Y, Z, by the content-giving terms, 'Socrates,' 'man,' and 'mortal.' It is easy to see that, by similar applications of the mathematical thinking that gave us (MP), we can obtain an endless number of mathematical propositions referring specifically to whatever variety of subject-matter you please. Let us make a few such applications. It will be illuminating to do so.

"If for brevity's sake we omit the logical primitives (a), (b) and (c) as being silently taken for granted, our very commonplace mathematical proposition (MP) may be stated with sufficient explicitness in either of the forms:

(MP) 'X is a Y' and 'all Y's are Z's' imply 'X is a Z,'

(MP) If X is a Y and all Y's are Z's, then X is a Z.

THREE MATHEMATICAL PROPOSITIONS IN ETHICS

"Now in (MP) let us replace

X by *the Golden Rule,*
Y by *moral maxim,*

and Z by *man-made empirical rule.*
The result is a mathematical proposition applied to the subject-matter of Ethics: If the Golden Rule is a moral maxim and all moral maxims are man-made empirical rules, then the Golden Rule is a man-made empirical rule.

"We may readily obtain another mathematico-ethical proposition by replacing

X by *the Golden Rule,*
Y by *moral maxim,*

and Z by *God-given rule.*

"And still another one by substituting for X and Y as above and letting Z stand for *insufficient guide.*

"Observe that all such propositions are mathematically sound notwithstanding the obvious fact that some of the propositions involved in them, being mutually contradictory, cannot all be true.

TWO MATHEMATICAL PROPOSITIONS IN THEOLOGY

"In like manner we may get a mathematical proposition relating to the subject-matter of 'fundamentalist' Theology. For that it suffices to replace

X by *Jesus,*

Y by *child of a virgin,*

and Z by *mythical being.*

"Another mathematico-theological proposition results from replacing

X by *Jehovah,*

Y by *brutal, ignorant, jealous, vengeful, abominable god,*

and Z by *being unworthy of human worship.*

A MATHEMATICAL PROPOSITION IN MYSTICISM

"Next replace

X by *what is received in ecstatic vision,*

Y by *revelation of ineffable truth,*

and Z by *alleged thing not within the range of scientific method.*

The result is a perfectly genuine mathematical proposition (which would be worth writing out in full) relating to the subject-matter of Mysticism.

A MATHEMATICAL PROPOSITION IN LAW

"For a final example replace

X by *an eternal principle of jurisprudence,*

Y by *thing revealed to the judicial mind by 'an overarching presence in the sky,'*

and Z by *non-existent thing.*

We thus obtain a beautiful mathematico-legal proposition, of much critical value in the philosophy of Law.

HELP YOUR NEIGHBORS TO UNDERSTAND

"It is evident that the sequence of such applications of just one little specimen (MP) of mathematical thinking admits of endless extension to and within every sort of subject-matter. But I believe that I have now performed the double task that your doubt and criticism imposed upon me a little while ago. For I believe you now see clearly that mathematical thinking is not essentially concerned with what is peculiar to any kind of subject-matter and that, being thus free, it is applicable to all kinds. And I trust that, when you hear people speak of mathematics as if it were the science of this or that particular sort of matter, you are prepared both to tell them, politely of course, and to show them that they do not understand. It goes without saying that, once they have listened attentively to you, they will understand."

ABOLISH THE TERMS PURE MATHEMATICS AND APPLIED MATHEMATICS

It is now manifest that what is called mathematical literature is comprised of works which fall into the one or the other of two mutually exclusive classes:[7] those which embody mathematical thinking but no application of it; and those in which such thinking and an application of it to some kind of subject-matter proceed simultaneously. In the latter case the sheer mathematical thinking and its application are often so fused as to give a superficial appearance of being but one thing. Does the foregoing classification accord with that which mathematicians generally have in mind when they speak in the familiar traditional way of 'pure' mathematics and 'applied' mathematics? It does not. Far from it. As com-

---

[7] Of course two works, one of the one sort and one of the other, may be and often are bound in one volume.

monly used, the term, 'pure' mathematics, covers not only such works as embody sheer mathematical thinking and nothing else but also many in which such thinking is applied to one or another kind of subject-matter; thus arithmetic, algebra, geometry, group theory, various function theories and numerous other doctrines are traditionally and commonly called branches of 'pure' mathematics, though all of them are applications of mathematical thinking to numbers or spacial entities or other variety of matter. On the other hand, the term, 'applied' mathematics, as commonly employed, does not cover the obvious applications I have just now indicated but covers only such as are found in mechanics, physics, chemistry, astronomy, statistics, and so on. It is plain that the two old terms in question have survived their usefulness. They serve only to blur and confuse, being false to fact, and they ought to be abolished. That is why in *Mole Philosophy and Other Essays* (p. 109) I have said:

"It is customary to speak of mathematics, of pure mathematics, and of applied mathematics, as if the first were a *genus* owning the other two as *species*. The custom is unfortunate because it is misleading. 'Pure' mathematics is a superfluous term, for it simply means (or ought to mean) mathematics and nothing else. The term 'applied' mathematics, which came into use before the essential nature of mathematics had been discovered, is a misnomer. The uses or applications of mathematics no more constitute a species of mathematics than the uses or applications of a spade constitute a species of spade."

WHY MATHEMATICIANS CONTINUE TO USE THE CONFUSING TERMS

Why is it that nearly all mathematicians persist in the world-befuddling use of such misleading terms instead of simply saying, and teaching the world to say, *mathematics* and mathematical *applications?* Partly no doubt because of the inertia of long-established habit but mainly, I believe, because of two additional facts. One of them is that the sharp line of distinction I have drawn between sheer mathematical thinking and such thinking fused with application of it is a recent discovery; the other fact is that the vast majority of mathematicians have meditated but little upon the question of the essential nature of their activity, are therefore but ill prepared to characterize it philosophically and so are not aware that they sometimes speak in a now inappropriate manner transmitted to them from the stone age of their subject.

ANOTHER ASPECT OF MATHEMATICAL THINKING

I have next to present an aspect of mathematical thinking which Pieri's formulation does not explicitly suggest. Before presenting it, it will be helpful as a preliminary to distinguish between having and seeming to have, between referring and seeming to refer. We have seen that the discourse of sheer mathematical thinking has in fact no kind of subject-matter, or content; yet there is a sense in which it *seems* to have. Let us notice what that sense is. Consider, for example, the proposition

(1) If a Y has two of its X's in a Z, then all of its X's are in the Z.

No meaning having been given to the marks

X, Y and Z, the proposition has no kind of subject-matter, or content. Yet the marks are present there and, when the proposition is uttered, they, too, are uttered by pen or by voice and are uttered *as if* they were content-giving terms, which they are not. Thus they lend to the proposition a certain semblance of content, though it has none. All this is obvious but it often happens that essentially the same kind of situation is much disguised. Consider, for example, the proposition

(2) If a line has two of its points in a plane, then all of its points are in the plane. Has it content? It has if and only if meanings have been assigned to the terms — point, line, and plane — for in themselves these terms have no more significance than have the mere marks X, Y, and Z. If, as often happens, proposition (2) occurs in a treatise in which meanings have not been given to the terms in question, then (2) is exactly on a par with (1); it has no content, or subject-matter, though, because of the terms, it seems to have.

It is precisely so in all strictly mathematical discourse. In any well-wrought mathematical treatise in which the thinking is not fused with applications of it but is sheerly mathematical there occur certain fundamental terms, as a rule only a few of them, which lend the discourse the appearance of being concerned with a kind of subject-matter, which it is not. These fundamental terms are, of course, *not defined*, for, if they were, the discourse would have some kind of subject-matter, namely the kind indicated by the definitions. We are thus led to consider an aspect of mathematical thinking which, as I have already said, is not explicitly suggested in Pieri's formula. I refer to the rôle played, in mathematical thinking, by Definition — the rôle of terms defined and especially of terms undefined.

The reader knows that mathematical works are distinguished by the large number of carefully defined terms occurring in them. And many people have the impression that in a well-constructed mathematical treatise its important terms are *all* of them defined. Of course they are not. No discourse, mathematical or non-mathematical, human or divine (whatever this may mean), can define all of its important terms. The reason is plain: there is no way to define a term except by means of other terms; and so, if we define certain terms by means of others, then these by still others, and so on, in the hope of defining all of our terms, we are bound to use, sooner or later, directly or indirectly, the terms first defined as means for defining others; and so our behavior will resemble that of a kitten pursuing its tail — a charming motion but no journey. Suppose one were to say: "Since I cannot define all of my terms, I will define none of them." The obvious answer would be: try the experiment and you will quickly find that your thinking cannot go far, that you are incapable of either producing or understanding scientific discourse, that your speech can rise but little above the level of mere chattering.

UNDEFINED TERMS OCCUR IN ALL DISCOURSE

It follows that every discourse necessarily contains some terms that it does not define. It follows, too, that a work consisting of *reasoned* discourse — by which I mean discourse primarily addressed to the logical understanding and so not merely or mainly designed to engender emotion—necessarily contains terms that it does not define and terms that it does define. In this respect what is the difference between an average or ordinary specimen of reasoned discourse, in physics, say, or ethics or theology or another field, and a well-built mathematical treatise? It is that in the latter the terms employed without defining them have been consciously selected for the purpose with the utmost deliberation and that all the other important terms have been defined with the same deliberation.

In a well-built mathematical treatise what is the relation between its defined and its undefined terms? The answer is that any defined term is either defined immediately by means of the undefined terms or else by means of terms that have been thus immediately defined or else by terms defined by terms defined by terms thus immediately defined, and so on, so that all the definitions rest ultimately upon the stock of undefined terms. And thus in such a treatise there is a striking analogy between the postulates and the propositions immediately or mediately *deduced* from them, on the one hand, and, on the other, the undefined terms and the terms immediately or mediately *defined* by means of them.

It is important to note that the undefined terms may be denoted, and in some otherwise well-made works unfortunately are denoted, by familiar dictionary words, such as point, line, number, integer and the like, thus making it necessary for the reader of such a work to guard himself against supposing that the words in question are really not undefined but are necessarily to be taken in their dictionary senses, which they are not; or the undefined terms may be, often are, and, ideally, ought always to be, denoted by obviously empty 'symbols,' or mere marks, like X, Y, Z and the like, for these will not, like ordinary words, tempt the reader to think the discourse is about some specific sort of subject-matter, when in fact it is not.

WHERE THE UNDEFINED TERMS OCCUR

Where do the undefined terms make their appearance? In any well-made work they enter in the beginning. They appear first in the postulates. It is the undefined terms which, as we have seen, give these initial propositional forms the semblance of having content, which they have not. The terms in question then appear throughout the entire work in all of the deduced propositions. These, it is true, contain defined terms and thus appear to be propositions about them. But, as we have seen, the definitions of all these terms rest ultimately upon the undefined terms. Hence the deduced propositions are, like the postulates themselves, nothing but propositions about the undefined terms. In other words, they are, like the postulates, propositional forms, devoid of content.

A MOT OF BERTRAND RUSSELL CRITICISED

I am here reminded of an oft-cited saying of Mr. Bertrand Russell: "Mathematics is the science in which one never knows what one is talking about nor whether what one says is true." It is obvious that the first half of the *mot* is true, for what one is talking about is nothing — nothing, that is, in the sense of 'no definite sort of subject-matter.' It is just as obvious that the second half of the *mot* is not true, for what one says in mathematics is that such-and-such propositional forms, postulates, imply other such forms, deduced propositions; and every such assertion, made after deduction, is perfectly true. A juster *mot* would be: Sheer mathematics is the science in which one never thinks of a definite sort of subject-matter nor fails to know that what one asserts is true.

USE OF UNDEFINED TERMS EXEMPLIFIED

What I have said of the rôle of definition in mathematical thinking — of the rôle in it of defined and undefined terms — is, I believe, fairly clear but it is a bit abstract and general. Perhaps it can be made clearer, a little easier to grasp, by a concrete or specific example. This time I will not use Euclid's *Elements*, though I might do so, but will use a recent work by an eminent mathematician, David Hilbert. In the English translation (by Professor E. J. Townsend) the title is: *The Foundations of Geometry*. It ought to be said, in passing, that the title is somewhat misleading in two respects. One of them is that, strictly speaking, the work is essentially no more concerned with distinctly geometric subject-matter than with an endless number of other kinds, being in fact essentially concerned with no kind at all. The other respect is that, even if we try to regard it as dealing essentially with geometry, we see that it cannot be regarded as dealing with every type or variety of geometry and that, therefore, it does not deal, as the title leads one to suppose it does, with the foundations of geometry *in general*. But let the question of title pass.

And now the example. I will be brief for I have discussed it extensively elsewhere[8] and need not repeat myself. Hilbert rightly begins by laying down a system of postulates (called axioms by him) selected with quite extraordinary deliberation. Their number is about a score. I will not set them all down here for a few of the shorter ones will do for my present purpose. Consider these:

(1) Two points determine a line.

(2) Three points, if not in a same line, determine a plane.

(3) If two planes have one point in common, they have another.

[8] See Keyser's *Mathematical Philosophy*, E. P. Dutton & Co.

(4) Of any three points in a line, one, and but one, is between the other two.

These and the remaining postulates, not here stated, seem, at first view, to have content, or subject-matter, because of the presence in them of the terms — point, line, plane, and space. But Hilbert has deliberately refrained from explicitly assigning meanings to these terms, and the same is true of two other important terms, relation-denoting terms: between and congruent. So we see that the postulates, being devoid of content, are nothing but propositional forms. It is now evident that the terms — point, line, plane, space, between, and congruent — deliberately selected by Hilbert for use as *undefined* terms might as well, or better, be respectively replaced by innocent marks — x, y, z, w, r, and $\acute{r}$. The postulates would then read as follows:

(1) Two x's determine a y.

(2) Three x's, if not in a same y, determine a z.

(3) If two z's have one x in common, they have another.

(4) Of any three x's in a y, one, and only one, has the relation r to the other two.

And so on for the other postulates.

The propositions deduced from the postulates do indeed contain many *defined* terms such as segment, angle, circle, sphere, triangle, and so on and so on. But all of them are defined directly or indirectly by means of the undefined ones, point, line, and so on. And so the deduced propositions are, like the postulates, just propositional forms, ultimately involving the *undefined terms* only. If we replace these by x, y, z, etc., then our defined terms will be defined by means of the marks, x, y, z, etc. Thus, instead of defining a circle as such-and-such a set of *points*, we will define it (call it durkol or something else if you like) as such-and-such a set of x's. Then all of Hilbert's deduced propositions will discourse about nothing but x's, y's, z's, etc., and various defined combinations of them.

QUESTIONS OF A CRITIC AND THEIR ANSWERS

That is the example. I hope it has served the purpose for which I have given it. Even so, a reader may say: "Suppose Hilbert had

The Realm of Mathematics

actually denoted his undefined terms by x, y, z, and so on, instead of the familiar words, point, line, plane, and so on, what would have been the gain?" The gain would have been a very precious thing—clarity; for his brilliant work would then have *appeared* to be what, contrary to its present appearance, it actually is — a fine specimen of sheer mathematical thinking, a specimen, that is, devoid of subject-matter of any kind. "That I see clearly," the reader may reply, "but pray tell me what possible use can such matterless thinking be in a world that is chockfull of subject-matter painfully pressing in upon us from every side." Well, that *is* a challenge and deserves to be met. My response is this: Any specimen of such thinking serves to reveal, if we but contemplate it, a very wonderful thing in the nature of our Universe — in the joint nature, I mean, of Man and the World; for it reveals the fact that those living organisms which we call human beings are endowed with capacity for a strikingly unique and characteristic type of *behavior* — that behavior which consists in entering, so to say, the infinite matterless realm of pure propositional forms and there tracing out the subtle threads of implication which bind such forms into coherent systems of forms indissolubly; and in that revelation is revealed, in their purity and nakedness, both the possibility and the essential nature of that strange familiar kind of human activity which from time immemorial has awed and mystified nearly all mankind — the possibility and nature, I mean, of sheer mathematics and sheer mathematical thinking. It is a revelation to man of a certain native potential dignity characteristic of man — a dignity in no way dependent on any particular subject-matter of the world. "That, too, I see," perhaps a reader will say, "and I own it is truly marvelous when seen steadily in clear light; but I was speaking of the 'use' of sheer mathematical thinking — of use in a somewhat lower sense of the term than you supposed. What I desired to know is the instrumental service of such thinking in a world crowded with particular subject-matters with which we have to deal constantly."

ANSWER TO A GREAT QUESTION

That is a *great* question. To answer it is a main motive of this chapter. Since the question is concerned with the applicability of sheer mathematical thinking, I have in several connections already answered it in part. I have invited you to discriminate between such thinking and applications of it and have insisted upon the importance of your doing so. I have pointed out that such thinking and an application of it may go on simultaneously and often do so, as in algebra, for example, or geometry, where the simultaneous activities are commonly so blended or fused that the author is hardly aware of his doing two things at once. On the other hand, applications of sheer mathematical thinking may come after, long after, such thinking has been done. Take Hilbert's book, for one among many examples. I have elsewhere[9] shown both that and how it admits of application, not merely to the sub-

---

[9] *Mathematical Philosophy: Lectures for Educated Laymen.* In this work I have given the name, Doctrinal Function, to any mathematical treatise in which the undefined terms are not interpreted, reserving the name, doctrine, to denote the result of any such interpretation.

ject-matter of geometry, but to other varieties of subject-matter — an endless number of them. Of course, everyone knows, or at all events believes, that one can think mathematically about the special subject-matter of arithmetic or that of algebra or that of geometry or that of physics or that of astronomy or that of statistics and some other sorts. What is not known — what is not believed but is disbelieved — what it is, therefore, immeasurably important to show to mankind — is the fact that one can think mathematically about any and every sort of subject-matter to which one may be drawn by love or driven by the exigencies of life. The fact ought not to astonish, for all subject-matters are but fragments or constituents or aspects of one whole — the world, including man or, if you prefer, included in him. But the very statement of it does astonish and, as said, is well-nigh universally disbelieved. That disbelief is the *Carthago delenda* of scientific methodology, popular 'philosophy' and vulgar ignorance. The attacks which we have thus far made upon it have been but little more than preliminary skirmishes. We have noted that every attempted defense by means of alibi involves a genuine bit of mathematical thinking by counsel, by court and by jury; we have signalised the fact that, wherever and whenever human beings have drawn an inference logically, they have thought mathematically; we have seen that a certain quite ordinary mathematical proposition, (MP), though itself devoid of content, can readily be so interpreted, or applied, as to yield countless mathematical propositions of importance about biology, about ethics, about theology, about mysticism, and about law; and we have seen that wherever propositions are bound to others by a thread of implication, in no matter what field of thought, it is the peculiar function of mathematical thinking to trace it out. No doubt such considerations ought to have the effect of weakening somewhat the enemy's self-confidence. But to win a decisive victory, to compel surrender—to vanquish completely that centuries-old world-wide disbelief in the general availability of mathematical thinking — it is *necessary* to make an analytical attack penetrating to its heart; and the attack would be *sufficient*, too, were it not required, in such cases, to slay the slain.

THE CONCLUSIVE ARGUMENT

The conclusive argument need not be long nor hard to follow. Suppose that I engage you in serious conversation respecting some subject in which you are interested, it may be in ethics or economics or religion or jurisprudence or commerce or medicine or sociology or politics or education, no matter what. Sooner or later you will assert some proposition which you deem to be important and which, let us suppose, is important. Then the proposition contains one or more important terms. In order that I may understand what it is that you have asserted, I request you to tell me the meanings of those terms — the meanings that you intend them to have in the proposition. You respond by defining the terms and of course you define them by means of *other* terms, for there is no other way unless you behave like that handsome kitten pursuing its tail, which you won't do for, by hypothesis, our conversation is serious and so must not be allowed to dissolve in laughter. I then ask you what those "other" terms mean. You answer by means of yet other terms. I next enquire about these. You respond again. I continue the pursuit, thus inevitably driving you to a set of terms of which you will say to me: "These terms I do not define; I use them as a stock by means of which I define the other ones." To which I reply: "In doing that, you are both exercising a right sanctioned by fate and performing a logical obligation; but you should not fail to observe that, in thus selecting the terms you employ without defining them and in defining your other important terms by means of them, you are strictly engaged in part of what is essential to thinking mathematically." As you go on exploring the subject (no matter what it is) that we are conversing about you will do a great deal of such thinking about terms, for, as we are soon to see, the need for it will recur again and again. Yet such thinking about terms, though it is indispensable, is not the main part of the mathematical thinking that you will do in your subject. That fact, too, we are soon to see. Let us return to the proposition that you asserted a little while ago and that led to my question respecting the meanings of its terms.

I own that you have well performed the task of enabling me to know what it is that your proposition asserts. I now request you to justify the assertion. You respond by offering what you call a 'proof' of the proposition. Any such 'proof,' whether good or bad, sound or unsound, has the nature and the form of an argument. As you know, any argument in support of a given proposition, makes use of one or more *other* propositions. Your argument is an attempt to show me that these 'other' propositions imply the one asserted by you; your attempted 'proof' of your asserted proposition is nothing but an attempt to deduce it from the 'other' propositions employed in your argument. These 'other' propositions must be relevant, hence they must relate to the subject-matter of our conversation; they must be not less important than the proposition you are trying to 'prove' by them, hence they must contain important terms. Unless these are among your undefined terms or are definable by means of them, you will be under the necessity of enlarging your present stock of terms undefined. Let us suppose this business attended to, so that I am in position to know what it is that those 'other' propositions assert. I next request you to justify their assertion. Again your response is a resort to 'proof' — to argument — in which you employ yet other propositions. I say 'yet other' for, were you to use some proposition already employed, you would thus commit the unpardonable sin of logical circularity and our discussion would instantly explode in laughter. Assuming that the 'yet other' propositions have been made clear in the usual (and only) way, I demand that you justify your assertion of them. And so on. It is perfectly clear that you will be thus inevitably driven to a set of propositions of which you will say to me: "These propositions I do not attempt to 'prove'; I deliberately select and adopt them as *postulates* from which to *deduce* my other propositions." To which I reply: "In so doing you are both exercising a right sanctioned by fate and performing a logical obligation; but pray do not, on any account, fail to observe that, in thus selecting a set of propositions to serve you as postulates and in deducing your other propositions from them, you are strictly engaged in thinking mathematically — just as strictly so as if the subject of your thought were number or mass or space or time or motion or force instead of ethics or economics or religion or medicine or law or something else, no matter what."

A LOGICAL USE OF VIOLET RAYS

If, after pondering the foregoing considerations, some reader still believes that some kinds of subject-matter do not admit of being thought about mathematically, I am prepared not to be astonished. For a myth that from time immemorial has been universally accepted as unquestionably true can hardly be destroyed by a short exposure to light however intense. Time is essential, an open mind and much repetition. If there be such a reader as I have supposed, I will once more try to cure him. There can be no remedy but light; let us bring to bear, if we can, the violet rays. I will ask the skeptical reader to name a subject about which, he is convinced, it is impossible to think mathematically. If he declines to do so, the discussion must end; such a skeptic may, possibly, be saved by the good old Catholic doctrine of invincible ignorance but not otherwise. If he names a subject — denote it by S — I will demand to know the ground of his conviction regarding it. He must then proceed to discourse about S, laying down one or more propositions relating to it and involving important terms. As in the foregoing conversation I will, of course, require him to tell me the meanings of his terms and to justify his assertion of the propositions. It is perfectly obvious that he will be thus compelled to resort to the now familiar method and apparatus of terms undefined, of terms defined, of postulates, and of deduction. What is the upshot? It is that our skeptic is forced into thinking mathematically about S by his very effort to prove that such thinking about S is impossible!

DO NOT MISTAKE DESCRIPTION FOR DEFINITION

This chapter should not close until we have considered a certain fundamental question that can hardly have failed to arise in the mind of any keenly attentive reader. Such a reader may say:

"I am thoroughly convinced that any subject whatever admits of being thought about mathematically. You have drawn a sharp distinction between sheer mathematical thinking and the application of it to a kind of subject-matter. I see that the distinction is fundamental and is too clear to be denied. You have shown, and have stressed the fact, that not only sheer mathematical thinking but any application of it expresses itself in discourse that is found, upon examination, to be discourse about some set of undefined terms. I can see no escape from that conclusion. But here a puzzling question arises that has not yet been answered. The question is this: Since, in the case of an application, the resulting discourse is, ultimately, discourse about a set of *undefined* terms, how does a reader of the discourse know that it is dealing with some sort of subject-matter and what sort it is?"

My reply is:

"He knows it, not in virtue of any definition, but in virtue of some *description*. Your question is extremely important, for the answer to it, which I hope soon to justify, is essential to any sound theory of knowing or knowledge. To justify the answer it is necessary first to distinguish between defining and describing. For, though definition and description are very commonly confused with each other to the frequent serious damage of otherwise rational discourse, yet they are in fact radically, I had almost said infinitely, different things. How may that be shown? A simple example or two will help to make it evident. Consider these statements:

(1) An even number is *an integer divisible by* 2.

(2) An even number is *a number such that, if 1 be added to it, the sum is not an even number.*

(3) An even number is *a number such that, if 1 be subtracted from it, the remainder is not an even number.*

(4) An even number is *a number such that, if an even number be added to it, the sum is an even number.*

(5) An even number is *a number different from the sacred number* 7.

(6) An even number is *a number that in some respects is like* 4, 6, and 8.

(7) An even number is *a number of a kind that some mathematicians have thought much about.*

"The italicised part of (1) is a *definition* (of an even number). Why? Please note the answer: Because the *name* — even number — has been given to it and to nothing else.[10] So you see immediately that the italicised parts of the other statements are not definitions; but, the statements being true, the parts in question are *descriptions* — partial descriptions — of what the term — an even number — denotes. The service of definition is logical; that of description is psychological. Obviously such partial descriptions are endless in number. Indeed it seems probable that *complete* description of anything would involve, or be,

---

[10] Of course, as the reader knows, one writer may give a name to a certain idea and to nothing else, and another writer may give the *same* name to a different idea and to nothing else. The name then has two genuine definitions, one in the text of the one writer and one in that of the other. But this fact does not invalidate the essential point of the above discussion.

complete description of every other thing. If it be so, then one contention of the mystics is just — that all is each and each is all — in the sense indicated. In the light of what has been said it is evident that an immense majority of so-called dictionary 'definitions' are, in the rigorous scientific sense of the term, not definitions at all but are at best nothing but partial, or incomplete, descriptions.

"It is of the utmost importance to note that there is one theoretically infallible criterion for deciding whether a combination of words is a definition of a term or only a partial description of the term's meaning, when we know that the combination is one or the other. The test is this: In any rigorously concatenated discourse, wherever the term occurs, it can be replaced by its definition without alteration of the sense or introduction of ambiguity but the term cannot be thus replaced by any partial description of its meaning. I have said "rigorously concatenated," for discourse may be, and most of it is, so logically 'rotten' as to render the stated test quite impracticable.

"I proceed now to justify the answer given

at the beginning of this reply. The answer is: In reading a dissertation in which sheer mathematical thinking is applied to some sort of subject-matter the reader knows that subject-matter is involved and what kind it is, not in virtue of any definition, but in virtue of some description. I am going to resort to examples as sufficient for my purpose. The description in question may be explicit or implicit or both.

"In my first example it will be both. The example is the *Elements* of Euclid, which I choose because of its familiarity — at least the name is familiar. Chief among the undefined terms are the terms — point, line, and plane — which, being undefined and having no *inherent* meaning (for no terms have such meaning), might as well be replaced by any other 'names' or vocables or marks, as X, Y, and Z. Yet every reader of that mathematical discourse is made aware of the fact that it is concerned with a certain type of subject-matter, namely, the properties of configurations in space. How made thus aware? The answer is: By the fact that, though the basal terms of the discourse are strictly *undefined*, the author has given partial but sufficient *descriptions* of the type of things *he* had in mind when using those terms. And the descriptions, as I have said, are partly explicit and partly implicit. The explicit descriptions are these:

(1) A point is *that which has no part*.

(2) A line is *breadthless length*.

(3) A plane is *a surface which lies evenly with the lines on itself*.

I have to note here that Euclid himself called those descriptions definitions, which he would not do were he living today. The blunder was very unfortunate, producing centuries of confusion and misunderstanding, retarding the philosophy of mathematics, scientific methodology, and kenlore, or theory of knowledge. And it *was* a blunder, for, though Euclid *called* them definitions, he did not *use* them as such. They play no logical rôle whatever in his work, by which I mean that they are never used, though all definitions are used, in the processes of deduction or demonstration. Their use is, not logical, but purely psychological, merely serving, as I have said, to indicate the kind of subject-matter that Euclid was applying his mathematical thinking to.

"I have said that Euclid gave the same

indication by implicit description also. What I mean is that the figures he has drawn show to the eye the type of things (spacial entities) he had in mind when using the undefined terms — point, line, plane — or combinations thereof.

"I hardly need cite another example, though there are many hundreds that might be cited. I will merely refer, briefly, to Hilbert's book, mentioned before. Here, too, point, line, and plane are among the undefined terms. Unlike Euclid, Hilbert gives no *explicit* description of what, if anything, he has in mind when using the terms. The sole clue to the fact that he has somewhat in mind and what it is — the sole clue by which the reader knows that the author is applying his mathematical thinking to spacial entities instead of some other kind or no kind at all and that, therefore, his discourse is geometric — is found in the implicit description given in the figures he has drawn; but that is a sufficient clue."

A PARTIAL SUMMARY

One who has read the preceding pages of

this chapter understandingly is qualified to affirm the following closely related propositions among kindred ones not here set down:

Mathematical propositions assert that such-and-such propositional forms imply other such forms and they assert nothing else.

Sheer mathematical thinking is not concerned with what is peculiar to the subject-matter of any type thereof.

The validity of a mathematical proposition asserting an implication among propositional forms having content, or subject-matter (as well as form), is absolutely independent of such content, or subject-matter.

The validity of mathematical propositions is independent of the actual world — the world of existing subject-matters — is logically prior to it, and would remain unaffected were it to vanish from being.

Mathematical propositions, if true, are eternal verities.

No mathematical proposition has been, is, or will be, contradicted, or invalidated, by any true proposition regarding any subject-matter that has been or is or can be.

There never has been nor is nor ever can be an actual world containing a kind of subject-matter admitting of no application thereto of sheer mathematical thinking.

The major part of existing mathematical discourse expresses the fusion of sheer mathematical thinking with application of it to one kind or another of subject-matter but sheer mathematics has its essential lair in implications, not in applications.

It is the applications of sheer mathematical thinking, and not the intrinsic qualities of it, that have commended mathematics to the 'practical' world and have won for mathematical activity some measure of allegiance and support — a measure that will increase in proportion as the world is made to see that the applications admit of endless multiplication not only but of extension to all subjects.

The intrinsic qualities in sheer mathematics have been discerned by very few, by hardly more than one or so among a billion of men and women. Yet it is these qualities that have caused some to repute the science "divine." Mathematics is not divine, it is human. The intrinsic qualities of it do not betoken any "divinity" in man; what they reveal is simply the intellectually best of that which in man is characteristically human.

All discourse, whether poetry or prose, whether mathematical or non-mathematical, whether it deal with the actual world or with the world of possibilities, is, ultimately, discourse about one set or another of undefined terms. That fact is a fact regarding the essential nature of discourse. Hence it is a fact regarding the essential nature of thought and hence regarding the essential nature of Man.

A mathematical discourse, if it be concerned with some type of subject-matter, discloses the fact, not by any definition of the subject-matter, but by some description of it — by partial description, which may be explicit or implicit and is often both.

Neither the actual world nor any part of it can be described completely but any part of it and hence the world itself can be described partially and endlessly.

Any thinking, whether mathematical or not, about some kind of subject-matter is thinking about something that is ultimately indefinable but is endlessly describable.

The function of definition, in our thinking about the actual world, is to enable us to deal effectively with Indefinables, which, notwithstanding they are indefinables, yet, fortunately, admit of endless Description.[11]

Absolute precision, absolute clarity, absolute rigor, absolute indubitability are unattainable but as ideals they are ineffably precious.

Sheer mathematics is essentially and exclusively concerned with the world of the Possible. It is *applicable* to the actual world because the world of the Actual is part of the possible, else it could not be actual.

[11] For further discussion of "Description and Definition" see Keyser's *Mole Philosophy and Other Essays*. Dutton & Co.

PART II
THE REALM OF SCIENCE

THE REALM OF SCIENCE

STATEMENT OF AIM

IN THE preceding pages I have endeavored to answer the question: What is Mathematics? And in the accompanying discussion I have tried hard to make the meaning of the answer, not merely intelligible, but clear. I venture to hope that the given answer, understood in accord with the given elucidations of its terms, will be fairly acceptable, not indeed to *all* mathematicians and philosophers, for that would be an unreasonable expectation, but to such of them as have meditated much upon the nature of mathematical activity with a view to ascertaining what it is that renders mathematical thought distinctive.

It is the aim of the present chapter to do for the term, Science, what the foregoing chapter has sought to do for the term, Mathematics, with this difference: The main task of the former chapter was to tell what the term Mathematics now actually means, while that of the present chapter is to tell what, in my opinion, the term Science *ought* to mean. The former is a report; the latter, a proposal. The aim is to formulate and submit for consideration a definition of Science that shall be at once sufficiently definite, clear and convenient to enable one to employ the term without the baffling vaguenesses and ambiguities that everywhere infest traditional and current usage; a definition that will, moreover, make it possible to discriminate readily and confidently, in the realm of thought, that which is scientific from that which is not, and thus enable one to recognize or to identify science as such wherever it occurs and whatever the guise in which it may present itself. It will be helpful, I believe, to say a preliminary word respecting the importance of such a definitional undertaking.

THE TWO MAJOR FUNCTIONS OF SPEECH

The reader is aware that language has two major functions — functions which, though not independent, are distinct, and which together embrace all the operations and uses of our human speech. One of the functions may be roughly described as the passional or galvanic or emotive function of speech; it is the power of words to express or to engender feeling. The other function may be roughly characterized as the logical or symbolic or illuminative function of speech; it is the power of language as a means or instrument for the symbolization, organization and communication of ideas or thought. I have said that the two great functions, though they are distinct, are not independent. It is almost certain that both of them are in some measure present and jointly operative in all discourse, for it is highly improbable that there can be an occurrence of pure thought — thought attended by no feeling — or of pure feeling — feeling accompanied by no thought. Yet in almost any given specimen of discourse it will be found that one or the other of the two functions predominates, and, in many instances, predominates in a measure nearly amounting to the complete exclusion of the other function.

More conspicuously than elsewhere the emotive or galvanic function of language manifests itself in rhapsodical discourse, in poetry, in the ecstatic utterances of mystics, in the passionate deliverances of fanatical evangelists, in profanity, in the harangues of politicians, in national shibboleths, in the manifold frenzied literatures of fear and hate and love. On the other hand, for the clearest manifestations of the logical or symbolic or illuminative function of language we naturally turn to the calm discourse of sober reason, where, as in some parts of the literatures of philosophy and jurisprudence, for example, but most notably in the literatures of natural science and mathematics, the controlling aim is, not the expression or the generation of feeling, but enlightenment of the understanding and the supreme concern is soundness of thought and clarity of exposition.

Regarding the galvanic function of language I will not here add anything to what I have said; but of the other function I desire to say something more. Its operation is so ubiquitous, so constant and so familiar that the symbolic or logical function of language seldom becomes a subject of conscious and deliberate attention; yet it is hardly possible to overestimate the importance of its rôle in the life of our human kind. For without it, without the power to symbolize ideas by words, and judgments by propositions composed of words, without the power to organize thought by the concatenating agency of logically interlocked propositional forms, without the power to communicate thought by speech, what we call Civilization could not have been produced; and were humans suddenly deprived of that power, they would suddenly sink to the intellectual level of dumb animals; they would cease to be human and human civilization would quickly perish.

I now wish to direct attention to another aspect of the matter. Granting that the logical function of language is in itself of the highest importance, we may ask how well the actual languages of the world are equipped for the performance of that function. The answer is that, despite their manifold emotive excellence, existing languages, even the maturest and most refined of them, when judged as logical instruments are found to be so rude and coarse, so grossly defective, as to seem infinitely inferior to the ideal of what such an instrument should be. That fact can hardly astonish anyone who understands that existing human beings, together with all their institutions and instrumentalities including speech, have been slowly and painfully evolved from pre-human and sub-human ancestry in a world where achievement is difficult and genuine ideals — the great lures to excellence — are never attained.

I have said that the defects of even the best of existing languages regarded as logical instruments are many and gross. It is not my intention to discuss here the general subject thereby suggested. For my present purpose it will suffice to indicate three or four of the graver defects in question and then to signalize one of them with special emphasis because of its special bearing upon the stated aim of this chapter. It will be best to begin with a certain question.

CONCERNING THE CONCEPT OF A LOGICALLY PERFECT LANGUAGE

The question is this: What is involved in

the conception of a logically perfect language? The question, which is concerned with a genuine and therefore unattainable ideal, is grave and difficult. I will not attempt to answer it completely but only in so far as the purpose of this chapter requires an answer. The partial answer to be given depends upon the answer to another question, namely: What is the characteristic service that such a language would be expected to render? To this question the answer is: To symbolize thought, to organize thought, to communicate thought, *unambiguously*. Perfect clarity and uniqueness of meaning, avoidance of ambiguity, of vagueness, of indetermination — that is the desideratum that would be the test. In relation thereto it is essential to bear in mind two cardinal considerations. One of them is that the meaning[1] of any given discourse depends upon the individual meanings of the propositions composing it and that these meanings depend upon those of the words involved. The other consideration is that a logically perfect

[1] Parts of *The Meaning of Meaning* by Ogden and Richards are well worth an attentive reading.

language could not be such unless it were available for *every* subject of thought, however common or rare, simple or complex, concrete or abstract, near or remote, old or new.

SOME NECESSARY BUT INSUFFICIENT CONDITIONS FOR A LOGICALLY PERFECT LANGUAGE

In view of what has been said it is evident that the following conditions are among those which a language, to be logically perfect, would have to satisfy:

(1) Every object of thought, however simple or complex the object, has a name in the language.

(2) No object has two or more names in the language.

(3) Every name in the language denotes an object of thought.

(4) No name in the language denotes two or more different objects.

It is customary and convenient to speak, in appropriate connections, of *necessary* conditions and *sufficient* conditions. A set of conditions such that a language satisfying them would be logically perfect would be a set of sufficient conditions. The four conditions above stated do not constitute such a set. They are merely some of the necessary conditions — of the conditions, that is, which a language would necessarily satisfy if it were a logically perfect instrument.

Notwithstanding the fact that the four conditions in question are far from revealing all that is involved in the conception of a logically perfect language, yet they are sufficient to show clearly and impressively that the logical defects of any existing language are both many and gross. As a logical instrument the English language may be taken as a fair representative of the best languages of the world. I do not ask whether it is a logically perfect language, for the question would be foolish, but I ask whether it satisfies even the stated partial list of conditions necessary, though insufficient, for such perfection. The answer is that it satisfies none of them and not only fails to satisfy them but fails so egregiously as to seem more distinguished for its logical defects than for its logical merits.

SOME OF THE LOGICAL DEFECTS OF EXISTING LANGUAGES

Comparing the equipment of the English language (or any other) with the requirement of condition (1), we are obliged to own that there are countless objects of thought, among them many objects of great importance, for which the language contains no names, and that with the growth of human experience, the number of such nameless objects continually increases more and more rapidly. Upon this point the first chapter of *The Logic of Modern Physics* by Professor Bridgman is illuminating.

The English language fails to satisfy condition (2) because there is a vast number of objects for which the language contains two or more names.

A like comparison with condition (3) shows that the English language contains a host of words which, though their proper function is strictly *emotive*, yet wear the appearance of being *logical* symbols — names, that is, for objects of thought — but which are found upon analysis to denote or symbolize no such objects and so, despite their appearance, are, in the logical sense, not names at all.

Finally, our English language (like the other great languages) violates condition (4), violates it outrageously, as any one can see. For, among the genuine names, the object-denoting names, in the language, there is an immense multitude of names each of which denotes, not merely one object of thought, as required by the *ideal*, but two or more, and often many, different objects.

The imperfections which I have briefly indicated are merely the more obvious ones among the logical defects of existing languages. They are sufficient, however, to account fairly well for the fact, so keenly realized by every disciplined thinker, that it is almost or quite impossible for us either to organize our thought rigorously or to communicate it unambiguously; and thus we may begin to understand why it is that the 'reasoned' literature of the world is, for the most part, fumbling and disputatious. The fate that has fashioned the languages of man is no doubt a very great poet but, as a logician, it appears to have been fairly idiotic.

GENUINE IDEALS ARE UNATTAINABLE

It must be owned, indeed I have already said, that the concept of a logically perfect language is a concept of an unattainable ideal. Is it to be inferred therefrom that the concept is barren? To attempt such an inference would be infinitely stupid. For all genuine ideals are unattainable — not stations to be reached in the courses we may run but transcendent goals, limits, that is, that lie beyond, not to be arrived at but to be approached step by step in an endless sequence of approximations. Their service is beyond computation; it is the most precious thing in the world. For it is the service of dreams of what ought to be; it is the service of incentives and lures to excellence; it is the service of perfect standards for estimating the worths of actual things and actual achievements; it is the service of the inner light that shows the possibilities and ways of unending amelioration; it is the service that beckons and woos to the pursuit of the Best; it is the service that quickens, sustains, ennobles and guides our human toil. Unattainable ideals have made it possible to win the great triumphs of the human spirit. Pursuit of them is the proper vocation of man.[2]

Genuine ideals are a kind of spiritual stars; like the physical ones they differ in rank, in dignity, in glory. Of all ideals, that of a logically perfect language, though its appeal is far from being universally felt, is one of the greatest. The sketch I have given of it is very incomplete, rudimentary, aiming at suggestion rather than portrayal or full presentation. My reason for sketching it at all in this connection and for here stressing its importance is this: The chief aim of the present chapter is, as before said, to formulate and present a satisfactory definition for the term Science, and I have hoped that the reader might be led to view the enterprise as I do, not as a mere attempt to fix the meaning of a word, for ordinarily that is a small matter, but as an endeavor to do an important bit of critical work in response to the lure, and under the authority, of a great ideal.

[2] For further bearings of this thesis see Keyser's *The New Infinite and the Old Theology* (Yale University Press) and the

THE TERM SCIENCE HAS NO STANDARD MEANING

It will hardly be denied that the word — Science — is today one of the mightiest terms in speech. What does it mean or signify? What does it symbolize or denote? The questions are to be taken as one. Let it be said at once that, inasmuch as words have no inherent meanings, the meanings they have are assigned meanings; words denote what they are employed to denote, symbolize what they are employed to symbolize, signify what they are employed to signify; they are to be construed as having the sense, or senses, in which they are used, and as having no other. Hence our question is equivalent to this one: What is it that the term, Science, is actually employed to mean or signify or symbolize or denote? In other words: What is the sense — or what are the senses — in which the term is used?

Everyone knows that the senses in question are many; how many and what they all are no one knows. To make a fairly complete list of them with suitable elucidations

first essay in his *Mole Philosophy and Other Essays* (Dutton and Co.).

of all its items would involve an immense amount of unprofitable labor. To see that it is so, it is sufficient to note, somewhat attentively, the great variety of senses in which the term science occurs in such familiar and current phrases as the following: natural science; normative science; material science, mental science; deductive science, inductive science; exact science, experimental science, descriptive science, speculative science; the science of logic, the logic of science; the science of ethics, the ethics of science; the science of metaphysics, the metaphysics of science; the science of life, the life of science; Christian science; mathematical science; domestic science; the science of palmistry; the science of religion, the religion of science; the psychology of science, the science of psychology; the science of education; the science of history, the history of science; the economy of science, the science of economy; engineering science, legal science, medical science, the science of theology; political science; occult science; the science of man, the humanity of science; practical science, theoretical science; esoteric science; the science of morphology, the morphology of science; the Science of the Sacred Word; and so on and on.

THE TERM SCIENCE IS EMPLOYED IN COUNTLESS SENSES HAVING NOTHING IN COMMON

To discover, assemble, compare and report all of the various senses in which the word science is employed would be a task so arduous and so unimportant that no one, unless possibly some desperate candidate for the doctorate of philosophy, may be expected ever to undertake it. It is legitimate, however, and will be helpful to employ the fiction that the task has been actually performed. Let us accordingly suppose that we have before us the immense collection of all the meanings or senses in question and keep vividly in mind that the collection is a *class C* whose *members* are the many distinguishable *senses* in which the term science is actually used in discourse. We may designate the various senses respectively by $S_1, S_2, S_3, \ldots S_n$, where n, we may be sure, is large and the S's, many of which are exceedingly vague, have, as a totality of individuals, nothing whatever in common, an astounding fact to be shown a little later.

How may we now indicate what the term science denotes? Adhering to the above-stated principle, that words denote what they are employed to denote, we may do it by pointing to the class C and saying: "The S's — the senses composing C — these are the things that the term denotes." And what justifies the statement is the fact that, by hypothesis, any one of the senses in C is a sense in which the term is sometimes used and that no one uses the term in a sense not included in C. The S's of C being objects of thought, it is seen that the term science, since it is a name of each one of a large multitude of different objects, is an exceedingly atrocious instance of the kind of words that violate condition (4), which as we have seen, is essential to the ideal of a logically perfect language. But that is not all. The atrocity I have signalized is greatly aggravated by the fact, above stated, that the various objects denoted by the name science — the many senses, that is, in which the term is employed —

have absolutely nothing in common. That they have in fact no element in common is readily seen. It is sufficient to consider but two familiar instances, say Mathematical Science and Christian Science. Compare them in any essential respect, in respect of content, in respect of form, in respect of method, in respect of spirit or aim. What factor have they in common? It is obvious that the answer must be: none whatever.

WHAT ACCOUNTS FOR THE SCANDALOUS SITUATION

The scandalous situation which I have depicted is mainly due to two powerful influences — influences that have always tended to lower the level and obscure the meaning of any term whose significance has given it great dignity and elevation. I mean the influence of downright ignorance, on the one hand, and, on the other, the yet more baleful influence of that active species of cunning which contrives to palm off doctrinal trash (or some other spurious ware) by the simple device of giving it a label chosen from among terms of high distinction, and thereby prospers because of the truly enormous capacity of mankind for being duped by the Magic of Words. Ignorance and cunning, sheer ignorance and the ignoble variety of cunning which I have described, these two have always been the most prolific breeders of dictional confusion, the chief degraders of grave and significant terms, and they have operated in every field of thought from time immemorial. But the damage they have done to the term science, prostituting it to the uses of downright ignorance and vulgar cunning, depriving it of definitely recognizable significance by making it stand for a conglomerate mixture of many meanings, diverse, irreconcilable, and void of a common factor, is a comparatively recent thing. Nor has the damage been done in secret; it has been wrought openly, publicly, even in the presence of 'scientific' men themselves.

ARE SCIENTIFIC MEN TO BLAME?

The question arises: What have these men done meanwhile either to prevent or to undo the damage? What have the devotees of 'Science' been doing to define and standardize the meaning of the great term, to defend its integrity, to maintain its dignity? What have they done to prevent or to arrest the increasing dissipation of its significance? To halt the growing degradation of its use? It must be said that an immense majority of them have done but little or nothing at all. They have been engaged in a kind of activity which they have called 'scientific,' in the study and teaching of what they have called 'science,' in the prosecution of what they have called 'scientific' research, and their achievements have been admirable. But just what they have meant by the terms 'science' and 'scientific' they have not made clear even to themselves, much less to the public. They have not even been concerned to do it; they have rather been haughtily unconcerned, as if the doing of it were a kind of philosophic task unworthy of their high calling. And so their knowledge of what they have meant by the terms in question may be likened to that which enables a duck to discriminate water from land or a horse to find its own stall — a valuable sort of knowledge but a kind below the distinctively human level of precise verbal signs and definite propositions.

What I have said of an immense majority of 'scientific' men can not be said of all. Fortunately there has been a minority, a relatively small minority, including men of great eminence, who, unlike the vast majority of their 'scientific' colleagues, have not, consciously or unconsciously, humbly or haughtily, assumed that they could know, without serious critical reflection, precisely what they mean by the terms 'science' and 'scientific.' On the contrary, these men have individually asked themselves the question: What is it that strictly *characterizes* what I am accustomed to call 'science'? What is the quality or property or mark that serves to discriminate the kind of activity or thought or knowledge which I habitually describe as 'scientific' from every other kind of activity or thought or knowledge? I have said "individually asked" because the question has never been submitted to a commission or a congress of 'scientific' savants. Each of the askers has answered the question in his own terms and each of them has endeavored in his own way to render his answer intelligible to the general public of educated laymen. The various answers are to be found in occasional addresses or in essays or books designed, at least in part, for lay perusal. A collection and suitable digest of them would constitute a fair-sized volume or two of interesting and edifying literature. I am not about to review it at length. With portions of it the reader is doubtless familiar. For my purpose it will be sufficient to give a few typical specimens of it as a mere reminder and to add a general characterization of the whole.

FAIR SAMPLES OF THE ATTEMPTS TO DEFINE SCIENCE

Here, then are some typical examples taken at random from a pretty wide variety of sources, mainly 'scientific':

"Science is knowledge gained by systematic observation, experiment and reasoning" (Pasteur).

"Science is knowledge of things, ideal or substantial" (Ruskin).

"Science is knowledge, not of things, but of their relations" (Henri Poincaré).

"Science is knowledge coordinated, arranged, and systematized."

"Science is ordered knowledge of natural phenomena and of the relations between them" (Whetham).

"Science is organized knowledge" (Pearson).

"Natural science is the attempt to understand nature by means of exact concepts" (Riemann).

"Science is organized common sense" (Thomas Huxley).

"Science is the search for the reasons of things" (Havelock Ellis).

"Science is the process which makes knowledge" (Charles Singer).

"Science is an attempt to systematize our knowledge of the circumstances in which recognitions occur" (A. N. Whitehead).

"Science has for its aim to discover and describe the orders of sequence and of coexistence that occur in the world."

"Science is the knowledge of the laws of phenomena, whether they relate to mind or matter" (Joseph Henry).

The foregoing instances will serve fairly well to exemplify what distinguished devotees or representatives of 'science' have accomplished in the way of formulating a crisp definition of the term. Such formulations are usually accompanied by more or less helpful explanations of the words employed, and fairness requires that any formula be construed and evaluated in the light (or the darkness) of such explanations. No one, I believe, can examine the literature of such formulations somewhat critically without being thereby led to note the following facts:

NOTABLE FACTS ABOUT THE SET OF FORMULATIONS

(1) None of the formulations is, strictly speaking, a *definition* of 'science' but is at best a more or less trenchant partial *description* of it or some salient aspect of it.

(2) The formulations differ widely in point of view: some of them regard 'science' as an enterprise, some as a body of achieve-

ments, and some as both; some endeavor to define 'science' in terms of subject-matter, some in terms of process or method, some in terms of spirit or aim, some in terms of the kind of knowledge that 'scientific' research is said to yield.

(3) The formulations differ immensely in respect to scope: some of them are so broad as to include in the domain of 'science' not merely astronomy, physics, chemistry, biology, and the like, but also philosophy, ethics, mathematics, linguistics, history, politics, rhetoric, engineering, Christian science, law, religion, theology, mysticism, and other matters of human interest; some rigorously exclude one or more or even all of the matters in the latter list; and some are so vague as to leave the boundary of the 'scientific' domain quite indeterminate.

(4) None of the formulations is the work of any congress or representative commission of 'scientific' men.

(5) None of the formulations is generally recognized by men of 'science' as authoritatively fixing, or standardizing, the term's meaning.

(6) There exists no formulation by means of which an intelligent layman can confidently discriminate, in accord with the judgment or usage of 'scientific' experts, what is 'scientific' work or thought or discourse or knowledge from what is not.

STATEMENT OF THE DESIDERATUM

Such is the present plight of a great term, I do not now mean in the usage of the uninformed or the cunning, but in that of experts, in the usage of those whose special interest, special obligation and special prerogative it is to define its meaning clearly and to guard its dignity against vulgar abuse and degradation. How can the situation be improved? The desideratum is to *define* the term Science in such a way that the definition will satisfy the following two conditions:

(a) It must serve intelligent laymen as a criterion for discriminating what is scientific in thought, discourse, or knowledge from what is not.

(b) It must be such as will win the approval of 'scientific' experts provided they examine it critically, without prejudice.

The Realm of Science

Can such a definition be invented? After much meditation I am convinced that it can, and I believe that the formula which I am going to submit will be found to meet the requirements. I earnestly desire, though I hardly dare hope, that no one will either adopt it or reject it before he has considered it maturely with open mind.

A FEW IDEAS RECALLED

My proposal is to define the term Science, not in terms of any kind or kinds of subject-matter nor in terms of any kind or kinds of method, but in terms of a certain kind or type of *propositions*. For the sake of emphasis and the reader's convenience I will here repeat, at least substantially, a few of the things said about propositions in the earlier pages of the preceeding chapter.

One of the things is a certain assumption: *To assert that p implies q, where p and q denote propositions, is equivalent to asserting that q is logically deducible from p.*

I will next request the reader to recall two definitions of the utmost importance: *If a proposition, P, is such that to assert it is equiv-*

alent to asserting that a proposition q can be logically deduced from a proposition p or — what is tantamount — that p implies q, then P is a Hypothetical proposition; in the contrary case P is a Categorical proposition. It is necessary to remember, too, that these definitions differ essentially from the usual (logic-book and dictionary) definitions of the same terms; that in the present work the former definitions are employed uniformly, the latter ones never; and that, accordingly, as shown in the preceeding chapter, either a hypothetical proposition or a categorical one may or may not be in the if-then form.

Without propositions the dance of human life could not go on. Propositions are present in our play and in our work, when we are awake and when we dream. Many millions of them are uttered daily by the men, women and children of the world; think of the untold billions that have been uttered in course of all the ages since human speech began and of the untold billions that will be uttered in the course of future ages before human speech shall end. Propositions are of many, many kinds, but there are two grand divisions that together embrace them all: the hypothetical and the categorical. Every proposition that has been, is, or can be, belongs to the one or the other of the two types, and none belongs to both — the types do not overlap. Nothing can be more radical or more significant than the distinction between them; it is a distinction that inheres in the nature of thought, and though it may be overlooked or ignored, it cannot be obliterated; to disregard it is always to invite confusion and frequently disaster. The fact that the distinction is fundamentally important manifests itself clearly in two cardinal considerations. Men habitually say that such-and-such a proposition has been established. What does "established" signify here? Applied to a hypothetical proposition, it means one thing; applied to a categorical proposition, the same word means an essentially different thing. For, as logicians are well aware, no hypothetical proposition can be established empirically — by the means, that is, of observation and experiment — nor by other means save that of deduction; and no categorical proposition can be established by deduction nor by any but empirical means. Again, men habitually say that such-and-such a proposition is true (or false). What does the adjective signify? As pointed out in the preceeding chapter, the word true (or false) has one meaning when applied to a hypothetical proposition and an essentially different meaning when applied to one that is categorical. To say that a hypothetical proposition, p implies q, is true (or false) is to say that p and q are (or are not) formally so related that q is logically deducible from p; but to say that a categorical proposition is true (or false) is to say that it states (or does not state) a fact verifiable empirically.

THE TWO GREAT CONCERNS OF THE HUMAN INTELLECT

Now, it is plain that concern with hypothetical propositions and concern with categorical propositions are the two great concerns of the human intellect. It seems perfectly evident that such infinitely momentous matters ought each of them to be designated by an appropriate name of its own; and it is equally evident that the two names, since they would properly symbolize disparate things ought in scholarly usage to be uniformly employed with strict regard to their respective significations. As indicated in the preceding chapter, the term Mathematics has recently come to be regarded by the best critics as the appropriate designation of the former matter; for the latter, which has never had a name, I venture to propose the term Science.

FORMAL PROPOSAL OF A DEFINITION OF SCIENCE

I submit the proposal in the form of the following definitions (of which the first three, concerning mathematics, are virtual repetitions):

A Mathematical proposition is a Hypothetical proposition that has been established.

Regarded as an enterprise Mathematics is characterized by its aim, which is that of establishing Hypothetical propositions.

Regarded as a body of propositions Mathematics consists of all and only such Hypothetical propositions as have been established.

A Scientific proposition is a Categorical proposition that has been established.

Regarded as an enterprise Science is characterized by its aim, which is that of establishing Categorical propositions.

Regarded as a body of propositions Science consists of all and only such Categorical propositions as have been established.

That is, in brief, the proposal. More precisely the proposal consists of the last three definitions or, more strictly, of any one of them for the three are virtually equivalent and may be employed interchangeably according to varying convenience. As indicated in the preceeding chapter the first three definitions are in substantial accord with what is now the best critical usage; they are, therefore, not "proposed" but are merely recorded here because of their helpfulness in considering what I have proposed. For it is evident that the established definition of mathematics and the proposed definition of science present the two great enterprises as coordinate parts in the composition of one immense picture — that of the whole effort of our human kind to understand the World including Man.

THE PROPOSAL EXAMINED FOR ITS PROS AND CONS

Two tasks remain: To consider what may be said in behalf of the proposal and what may be said against it.

The former task has been performed in part. We have seen that a proposition is either hypothetical or categorical and is never both; that, accordingly, the world of propositions presents two grand divisions or realms, one composed of hypothetical propositions, the other of categorical ones, and that the two realms have no proposition in common; we have seen that the adjective true (or false) has two essentially different meanings according as it is applied to a proposition of the one realm or to one of the other; we have seen that a true proposition of the one realm and one of the other cannot be established by the same means. We have observed that the truth-seeking activity of the human intellect is accordingly composed of two distinct enterprises differentiated and characterized by their respective aims; that the aim of one of them is to establish hypothetical propositions, and that the aim of the other is to establish categorical ones. We have agreed, I think, that these coordinate enterprises, the two great concerns of the human intellect, ought each of them to be designated by a name of its own. We have noted that the former enterprise has such a name, Mathematics, in virtue of which all mathematical propositions, however much they may differ among themselves, are known to have one mark in common — they are all of them hypothetical. We have noted at the same time the astonishing fact that the other enterprise, than which nothing in the world is more important, has never received a name. My proposal is to call it Science.

What advantages might be reasonably expected to occrue from general adoption of the proposal? Some of them are evident immediately. One of them would be the very great convenience of having a recognized name for a hitherto nameless enterprise of universal interest and the greatest possible moment. Another advantage would be that, as in the case of the mathematical propositions, all scientific propositions, however much they might differ among themselves, would be known to have one character in common — they would be, without exception, categorical. At present 'scientific' propositions have no character whatever in common. A tremendous advantage would be that of bringing to an end the nearly universal and immeasurably injurious confounding of the mathematically true (or false) with the scientifically true (or false); and the like is to be said of a kindred confusion — the confusion, I mean, of the mathematically 'established' with the scientifically 'established.' An additional advantage of no little importance would be that of enabling us to give a clear and definite answer to an endlessly debated question of 'scientific' methodology. I mean the question: What is the 'scientific' method? The answer would be: The scientific method consists of any and all available means for the establishment of categorical propositions. Thus conceived, *the* scientific method would obviously embrace all special scientific methods whether known or yet to be discovered, but none of these could rightly claim to be *the* method of science. Among the advantages that would accrue I have now to signalize one which is undoubtedly very important and may indeed be, in a sense, the greatest of them all.

I have said that the desideratum is to define the term Science in such a way that the definition will satisfy two conditions. One of these is that the definition shall serve intelligent laymen as a criterion for discriminating what is scientific in thought, discourse, or knowledge from what is not. Would the proposed definition, if adopted, fulfil that requirement? I venture to assert that it would. In doing so I make two assumptions which I believe will be regarded as fair. One of them is that an intelligent layman can readily learn to tell whether any given proposition is hypothetical or categorical in accord with the meanings of the terms as herein defined. The other assumption is that an intelligent layman will know how to ascertain whether a given proposition is or is not an "established" one. It is important to say here what we mean by an "established" proposition. We do not mean a proposition that is absolutely known to be true or one that cannot be doubted. The question is sheerly a question of right usage, and I believe that the answer will be found, upon reflection, to be this: An "established" proposition is one that is so regarded, so treated, so spoken of by all or nearly all expert authorities in the field or the subject to which the proposition belongs. It is plain that the element of time is essential. An established proposition has a date or dates. The history of thought makes it abundantly evident that a given proposition may be, for a period of time, an established proposition and then cease to be such. The Newtonian law of gravitation; the nebular hypothesis of Kant and Laplace; Jesus was born of a virgin; the earth is flat; every continuous curve admits a tangent at each of its points; any whole is greater than any one of its parts; heat, light and electricity are imponderable substances: these are but a few of many propositions that were once established propositions but now are not. So it is seen that the second of the foregoing assumptions amounts to no more than this: that an intelligent layman, if he does not know, will have gumption enough to ascertain, by the simple device of asking some expert authority, whether, at the time of the inquiry, a given proposition is or is not an established one.

Suppose that such a layman resolved to examine, in accord with the proposed definition of science, some ostensibly reasoned discourse — some seemingly important work on no matter what subject — with a view to discovering for himself what, if any, portions of the work are scientific; what, if any, portions of it are mathematical; and what, if any, portions are neither the one nor the other. What would be his procedure? In outline it would be this: Confining his attention to the theses of the work, he would ascertain, in the way above indicated, which, if any, of these are established propositions, and which, if any, of them are not. In the case of the established propositions, if there be any, he would note which, if any, of them are hypothetical and which, if any, are categorical; he would say that the former propositions are mathematical and that the latter are scientific; if he found that some of the theses are not established propositions, he would say that, though they are propositions, they are neither mathematical nor scientific propositions.

Moreover, regarding any truth-seeking activity, his own or that of another, that of an individual or that of a group or an institution, our layman would be enabled to infer from its aim whether, as an enterprise, it is scientific or not and whether it is or is not mathematical; he could know, that is, to which of the two great hemispheres of intellectual life the activity in question belongs. And it will hardly be disputed that to deny high value to such knowledge, to depreciate such orientation in the geography of thought, would argue an unworthy conception of the proper dignity of man.

We must not lose sight of the obvious fact that no definition, whatever its intrinsic merits, can be effective unless those for whose use it is specially designed become willing to employ it. And so I have said that a definition of Science ought to "be such as will win the approval of 'scientific' experts provided they examine it critically, without prejudice." I have briefly indicated some of the advantages which might, I think, be confidently expected to result from adopting the definition proposed. It will hardly be denied that they constitute a somewhat impressive array. Are they sufficient to win the required approval? That question cannot be answered without considering carefully such responses as a typical 'scientific' expert is likely to make to the following questions: What is the certainty and what the value of the alleged advantages? And, supposing that such advantages would actually accrue, would the gain of them be attended by serious disadvantages — by a corresponding loss, that is, of important advantages now existing?

WHAT A TYPICAL 'SCIENTIFIC' MAN IS LIKELY TO SAY OF THE PROPOSAL AFTER SOME MEDITATION

Let us now try to imagine what our typical expert will probably say in the premises. He is required, by hypothesis, to examine the proposed definition "critically, without prejudice." On this account he may be fairly supposed, I believe, to begin as follows:

"After a good deal of reflection I approve as fundamental the way of your approach on account of its primary regard for the nature of propositions and I find myself in accord with much of what you have said. It is too evident for denial that any proposition whatever belongs to one or the other of two mutually exclusive classes, or realms, the realm of hypothetical propositions and the realm of categorical ones. I agree that in good usage the term, "established" proposition, signifies what you have indicated. I agree that a hypothetical proposition cannot be established empirically nor a categorical one deductively. I agree that the word true (or false) has one sense in the hypothetical realm and an essentially different sense in the categorical one. In view of such basic considerations it is perfectly evident, and I entirely agree, that truth-seeking activity comprises, as you have said, two distinct and coordinate enterprises which are characterized by their respective aims and together embrace the intellectual life of man. No one will deny that each of the great enterprises ought to be everywhere designated by an appropriate name. I am glad to be definitely and clearly informed, for I had not been fully aware, that that one of the enterprises which has for its aim to establish hypothetical propositions is now denoted in the best critical usage by the term Mathematics. And, of course, I grant that it would contribute greatly to the convenience and clarity of human discourse, could we find a suitable designation for the complementary enterprise, which has for its aim the establishment of categorical propositions. But, as you know, my granting of this does not imply approval by me of your proposal to designate the enterprise in question by the term Science. And, unless I am mistaken, there are some good reasons for withholding such approval. I wish to state them and will begin with one that seems to be pretty obvious.

MATHEMATICS OUTSIDE THE DOMAIN OF SCIENCE

"It is perfectly evident that, were the proposal adopted, one of the immediate effects would be complete exclusion of mathematics from the domain of science. That fact is alone sufficient to show how very revolutionary the proposal really is. For to contemplate the exclusion of mathematics from the domain of science is to contemplate shocking violence to one of the best established and most venerable of traditions. From time immemorial, as you know, mathematics has been universally regarded by scholars, including mathematicians and 'scientific' men, not only as being a branch of 'science' but as being, among all 'scientific' branches, the one best entitled to be called 'scientific'; the view has been that mathematical knowledge is 'scientific' knowledge *par excellence*; not only have many 'scientific' propositions been mathematical but all mathematical propositions have been 'scientific.' But your proposal, if adopted, would terminate the age-long career of that traditional view, for then we should have to say that no mathematical proposition is scientific and no scientific proposition is mathematical. That is why I have said that the proposal is shocking.

BUT TRADITION IS NOT INTRINSICALLY SACRED

"I trust you are not tempted to infer or to imagine that I am capable of defending a tradition merely because it is a tradition. Neither as a 'scientific' man nor as a representative of such men could I possibly do that. I know well that many a hoary tradition, having originated in erroneous views and having been perpetuated by the sheer force of inertia, has had at length to be condemned and abandoned in the interest of progress. It may even be that all traditions, however seemingly well established, are destined sooner or later to meet such a fate. However that may be, I should be at present unwilling to abandon the particular tradition which I have cited, for I believe that, respecting the inclusion of mathematics within the domain of 'science,' the tradition is sound. I have now to give my reasons for thinking so.

THE 'SCIENTIFIC' MAN LOVES TO CLAIM MATHEMATICS AS THE IDEAL BRANCH OF SCIENCE

"One cannot glance at 'scientific' literature

— at literature, I mean, which you will not deny is scientific — without perceiving that many of the propositions found there are evidently mathematical. It so happens that I am myself a physicist, and I assume that your proposal would not exclude Physics from the domain of science. Well, in the literature of physics mathematical propositions abound, so much so that there are many physical treatises and memoirs that cannot be read understandingly even by physicists unless they are at the same time able mathematicians. I need not remind you of such works as Newton's Philosophiae Naturalis Principia Mathematica, Lagrange's Mécanique Analytique, Maxwell's Electricity and Magnetism, or Einstein's Theory of Relativity, among a rapidly increasing number of like kind. What I have said of the literature of physics may be said of that of other 'scientific' subjects, of chemistry, for example, or botany, or psychology, or statistics, or economics, and so on. It appears that mathematics is not merely a branch of 'science' but it is the model branch thereof — an ideal pattern to which other branches, as they approach maturity, more and more conform.

"Moreover, what I have been just now saying seems, somewhat strangely, to be strongly confirmed by some of the main things that you yourself said, and said with much emphasis, in the preceding chapter. Let me specify.

"You quoted with approval, and carefully explained, a certain neat definition of mathematics which explicitly represents mathematics as a branch of science. That definition states, quite consistently with the traditional view, that mathematics is "the hypothetico-deductive science." It is true that what you there stressed is the "hypothetico-deductive" character of mathematical thinking but you were silent regarding any impropriety in the definition's employment of the term "science."

"Again, in that chapter you were much concerned to show, and I think you did show conclusively, that mathematical thinking is available for any kind whatever of subject-matter. If, as you assert, it is possible to think mathematically about any given subject, then it must be possible to establish mathematical propositions relating to any given subject that you would call scientific. It seems to me natural to say that such mathematical propositions, since they relate to a scientific subject, are also scientific propositions. Suppose, for example, that I, being a physicist, establish a mathematical proposition in the field of physics, which you will not deny is a scientific field. What good ground have you for maintaining that that proposition, though it is mathematical, is not also a physics proposition and hence a scientific one?

ARE HYPOTHESES TO BE THE EXCLUSIVE CONCERN OF MATHEMATICS?

"But that is not all. I am puzzled in trying to reconcile another of your contentions with your proposed definition of science. You seem to hold that *hypotheses* are the peculiar or exclusive concern of mathematics. You tell us that mathematics is the enterprise which aims to establish hypothetical propositions and then you propose to define science in such a way that mathematics can no longer be regarded as a branch of science. Is it true, as it seems to be, that, according to your conception of science, scientific men cannot, as such, employ hypotheses or speak of a working hypothesis or of hypothesis and verification as an essential part of scientific method? You see why I am puzzeld.

THE 'SCIENTIFIC' EXPERT SUGGESTS AN ALTERNATIVE PROPOSAL

"In view of such considerations it seems to me that, instead of defining science as the enterprise having for its aim the establishment of categorical propositions, it would be far better to find some appropriate name for that categorical enterprise (as we may call it for short) and to reserve the term Science to cover at once both the Categorical enterprise and the Hypothetical one (or Mathematics). By what name the categorical enterprise ought to be designated I am not now prepared to say, but I have no doubt that a suitable name could be found or invented. If it were done and the term science were employed as I have suggested, it is evident that the following would be among the resulting advantages: each of the great coordinate enterprises would have an appropriate name of its own; we should have a name for the two combined; and the venerable tradition, according to which mathematics is a division of science, would be preserved."

Such I believe to be a fair representation of what a typical 'scientific' expert is likely to say regarding the matters in question. He may wish to say more. Before giving him an opportunity to do so it will be best, I believe, to consider what he has already said, for thus the discussion will be easier to understand. My response, then, is as follows:

RESPONSE TO THE 'SCIENTIFIC' EXPERT'S ANIMADVERSIONS

"I note with much pleasure your agreement with what I have said respecting the basic rôle of propositions in such a critique; respecting the two propositional realms; respecting the two meanings of the word true (or false); respecting the two meanings of 'established'; respecting the right use of the term, established proposition; respecting the two great component enterprises of truth-seeking activity; respecting the designation of one of them by the term mathematics; and respecting the desideratum of finding or devising an appropriate designation for the other one. In such fundamental points of agreement I find ground for hoping that you may yet approve my proposal despite your present animadversions against it. In dealing with these I need not follow the order in which you presented them.

"First of all a word as to my having (in the preceding chapter) quoted, with warm approval, a definition that expressly represents mathematics as a branch of science. I did not there raise any question regarding the propriety of such representation. I deliberately refrained from doing so and the reason was this: I desired to concentrate attention upon the "hypothetico-deductive" character of mathematics or mathematical thinking and was well aware that to broach the other matter at the same time would both divide the reader's attention and further complicate a complicate discussion. That is why it was reserved for consideration in the present chapter.

THE PROPOSED DEFINITION OF SCIENCE LEAVES UNIMPAIRED THE RÔLE OF 'SCIENTIFIC' HYPOTHESIS

"You have intimated that my proposed

definition of science is inconsistent with the employment of hypotheses in scientific research, and you have challenged me to show that such is not the case. I am glad to accept the challenge. I deny that my proposal would restrict the use of hypothesis to mathematics. I assert that under the proposal scientific men could employ hypotheses as freely as they have ever done. To justify the assertion it is essential to remember that, as I have said, the scientific method would "consist of any and all available means for the establishment of categorical propositions." Is logical deduction among the available means? You and I agree that no categorical proposition can be established by deduction alone; we agree that for the establishment of such a proposition one has in every case to depend ultimately upon empirical evidence — upon the witness, that is, of observation and experiment; but, though deduction alone is never a sufficient means to such establishment, it is often a very powerful auxiliary means thereto — a means for discovering the kind of evidence that *is* sufficient; upon that, too, we doubtless agree. Hence we agree that the answer to the question is yes: under the proposed definition of science logical deduction is a proper part of scientific method — a potent means, that is, though never in itself a final or sufficient means, for establishing categorical propositions. Now, such deduction by scientific men deals with what you at present call, and under the proposal will continue to be called, hypotheses — scientific hypotheses. And so are completely justified both my denial that the proposal would restrict the use of hypothesis to mathematics and my assertion that under the proposal scientific men could employ hypotheses as freely as ever they have been wont to do.

"You have directed your criticisms against that feature of the proposal by virtue of which mathematics would be excluded from the domain of science, so that — contrary to the traditional view, which you maintain — no mathematical proposition would be scientific and no scientific proposition would be mathematical. This you find shocking. One of your criticisms of it charges me with self-contradiction. You quote and heartily endorse my saying that it is possible to think mathematically about any subject whatever and then you allege that to say that is incompatible with saying at the same time that a mathematical proposition ought not to be called scientific even when it relates to the subject-matter of a well-recognized branch, say physics, of empirical knowledge. The allegation has a plausible aspect. In order rightly to examine the merits of the case I will return to the matter of hypotheses. For it is essential to observe carefully that the rôle of hypothesis in mathematics is one thing, that its rôle in physics, for example, or astronomy or biology and so on is another, and that the difference between the two rôles is radical. The difference is a difference of aim and of method, and to avoid confusion we must keep it constantly in mind.

"In indicating the difference in question I will specifically refer, for the sake of a little concreteness, to physics, since you are a physicist, for it will be evident that what I am going to say applies equally to all branches of empirical knowledge.

HOW AND WHEN MATHEMATICS INTERVENES IN 'SCIENTIFIC' RESEARCH

"I will begin by asking how you as a physicist come to be sometimes interested in mathematical propositions or in the deductive process by which they are established. You will agree, I am confident, that the answer is substantially as follows: You observe a certain group of seemingly related physical phenomena; you desire to account for them, to ascertain their cause or the necessary and sufficient conditions for their appearance or perhaps the law to which they conform (for you assume that there is such a law). If known, the cause or the conditions or the law would be stated categorically as a fact regarding the phenomena in question; and so it is evident that your aim is to establish a categorical proposition about a part or an aspect of the physical world. You make a more or less shrewd guess as to what the sought-for proposition is; the guess requires to be tested; in the process of testing it the first step is to ascertain the consequences or implicates of the guessed-at proposition; these are also categorical propositions (true or false) regarding the physical world; the next step in the testing process is to determine whether or not these implied categoricals are valid statements of physical fact including the phenomena to be explained, and this you do by means of observation or experiment or both. If you find that at least one of the phenomena remains unaccounted for or that at least one of the implicates in question is invalid, you immediately reject your guessed-at categorical as false and guess again; but, if all of its implicates are found to be valid and to cover the phenomena in question, it thereupon acquires the status of an established proposition. What I have given here is merely a rough outline of your procedure but it is sufficient for my purpose. It is easy to see at what stage of it you are obliged as a physicist to invoke the help of mathematics: it is the stage in which you have to ascertan the logical consequences of your guessed-at proposition. It may be that you will deduce the consequences your self, for it sometimes happens that a physicist is a mathematician also, but, for the sake of vividness and to avoid the possibility of confusing the two characters of a double personality, I will suppose you to call in a mathematician to make the deductions for you. Grant that this has been done and that your mathematician has made the required deductions. Denoting your guessed-at proposition by p, and the set of propositions deduced therefrom by q, he hands you for your use the following mathematical proposition

(M) p implies q,

which is often, though not quite equivalently, expressed in the familiar form

(M) If p, then q.

"I now invite you to note very carefully that "radical difference" which I alluded to a little while ago and promised to indicate — the difference of interest and method and aim. It is seen in the following facts:

REVEALING CONTRASTS BETWEEN THE CONCERNS OF THE MATHEMATICIAN AND THOSE OF THE MAN OF 'SCIENCE'

"Your mathematician's interest began and ended in the proposition (M); it began with the question — what propositions are logically deducible from p? — and ended when the answer was found. But your own interest neither so begins nor so ends; it begins with a certain group of physical phenomena of which you desire to discover the cause or the conditions or the law and it does not end until the discovery has been made and has been expressed as a categorical proposition, asserting that such-and-such is the case.

"Your mathematician's aim was to establish a proposition (M), which is hypothetical. But your aim is to establish a proposition p, which is categorical.

"For both you and your mathematician, p is an hypothesis; but for the mathematician the hypothesis is merely the implier of q; for you, however, as physicist, its rôle is radically different (and ought to have another name), for what it represents is precisely your to-be-tested guess at the cause or conditions or law you are trying to discover. In the former sense, the purely logical sense, p is an eternal hypothesis, as it implies q eternally; but in the other sense it is not, for p may be rejected as false or it may become an established proposition, and in either event it ceases to be an hypothesis in your sense of the term, in the sense, that is, of the familiar phrase 'scientific hypothesis.' What this phrase now signifies, and under my definition of science would continue to signify, is nothing but a categorical proposition conjectured to be true and submitted to examination to ascertain whether or not it actually is true.

"Your mathematician is not at all concerned with the truth or falsehood of p and q, for his proposition (M) merely asserts the deducibility of q from p, and such deducibility depends exclusively upon the *forms* of p and q and not upon their content, or what they say; but as a physicist you are so essentially concerned with their content, with what they say, with their truth or falsehood, that, if you discover either p or q to be false, you immediately cease to have any interest whatever in (M), notwithstanding the fact that (M) is true.

"What your mathematician asserts is (M); but what you, as a physicist, hope to be able to assert is p or q or both of them; and his assertion neither supports yours nor is supported by it.

"For your mathematician the proposition (M) is strictly building material, it is the very fibre and stuff of mathematics — an edifice of hyptheticals like (M); but for you, as physicist, (M) is not such material, it is strictly nothing but an instrument or tool used in constructing Physics — an edifice of categoricals like p and q."

It is, I think, highly improbable that our physicist, once he has considered them, will dissent from any of the sharply contrasting statements made in the last six paragraphs. But improbable events do sometimes occur. And so, to provide against the possibility of being surprised by an extremely improbable eventuality, I am going to suppose our friend to rise once more in defense of his old thesis that the mathematical proposition (M) is also a *physics* proposition. I will ask him to state precisely why he asserts it and will suppose his answer to be this: "I assert that the proposition (M), though it is mathematical, is also a physics proposition because it is a true proposition about the subject-matter of physics."

THE WORTHLESSNESS OF A BEST ANSWER

That being his answer, I will offer the following in reply:

"Your answer is, I think, the most plausible that could be made. Yet I am compelled to say that it is absolutely worthless and the compelling reason is that, were you to discover p or q to be false, you would, as above said, immediately reject (M) from consideration, notwithstanding the rejected (M) would perfectly satisfy your condition of being 'a true proposition about the subject-matter of physics.' If you will permit me to say so, you appear to have been confused and misled by the double meaning of the words 'about the subject-matter of physics.' Applied to p or q, they have one meaning; applied to (M), another. The proposition p, for example, is 'about the subject-matter of physics' in such a sense that, to ascertain whether p is true, you have to examine that subject-matter itself and compare it with what p asserts. But to ascertain whether (M) is true, you do not look into any physical subject-matter nor even into p and q but only at their *forms*, disregarding their content or physical significance completely.

AN APPEAL TO COMMON SENSE

"I am going to let my final word on this matter be an appeal to your common sense. Suppose you say to your servant: 'John, go and examine our neighbor's dog Fido and then come and describe him to me.' After a little time John returns to describe the dog. He will make categorical statements telling Fido's color, his size, how many legs he has, how many eyes, how many noses, how many tails, and so on. Now, if John were to add solemnly, *as part of the description of Fido,* the statement — if Fido has two ears and two eyes, then Fido has as many ears as eyes — or the statement — if Fido has ten tails and four noses, then Fido has more tails than noses — what would you think? I think you would think what I think when you say that (M) is part of the description of *your* Fido — the subject-matter of Physics.

THE UPSHOT

"What is the upshot? Since what I have said of physics is valid for any other branch of empirical knowledge, the upshot is this: mathematical propositions are available for use as instruments in physical or chemical or astronomical or biological research but they cannot be significantly said to be physics propositions or chemistry propositions or astronomy propositions or biology propositions, and so on for all the other subjects or divisions of what you call empirical science.

IT WOULD HAVE EVOKED LAUGHTER

"I cannot but wonder whether there remains in you any inclination to continue your defense of the old tradition that mathematics is a branch of science. If so, I desire to say an additional word regarding it. We agree in our recognition of the two great propositional enterprises — the hypothetical and the categorical. We agree that mathematics is the former one; you will not dispute that what you call empirical science is a part (at least) of the latter one. Now, as you are well aware, it so happens that in current usage the term science and the term empirical science are commonly identified and employed interchangeably. Consequently your tradition amounts, in practice, to saying that the hypothetical enterprise is a part of the categorical one; which is very confusing to the vast majority who do not perceive that it is a naked absurdity. I am not going to reiterate the differences between the two enterprises but will submit this proposition: The differences are so numerous and profound that, had they been known in the remote past when the tradition began, you would not now be defending it because it could never have begun; to have proposed it would have been to evoke laughter.

PRAGMATHETICS BUT THE SUGGESTION IS TOO LATE

"Thus far I have said nothing explicitly about your proposal or suggestion that we find or devise for the Categorical enterprise an appropriate name (other than Science, proposed by me) and reserve the term Science to designate at once both that enterprise and the Hypothetical one (or Mathematics). Your suggestion seems to me to be good logically but bad psychologically. Had we a clean slate in the matter I should be quite willing to denote the categorical enterprise by the term *Pragmathetics* (from *pragma* and *thesis*) for the designation would serve to remind us of the fact that an established categorical proposition rests ultimately upon evidence that is empirical, or *pragmatic*. Were that name adopted, a mathematical (or mathetic) proposition would be, as before, an established hypothetical one, and a pragmathetic proposition would be an established categorical one. But the slate is not clean. For a very large part of the territory that would thus belong to pragmathetics was long ago squatted upon, and has ever since been occupied by, what you frequently style Empirical Science. The tenant cannot now be dispossessed — for that it is, psychologically, too late. I believe that my proposal is more feasible, for what it says to the tenant is virtually this: 'You have long owned a very large part of the pragmathetic territory; keep it; at the same time have the discernment to see that the remainder of the territory, being categorical, empirical, pragmatic, is essentially the same in kind as the part you have long been cultivating and now possess; hence extend your claim to cover the whole and possess it all as yours by natural right; relinquish your absurd pretension to ownership of mathematical territory; stop your confusing practice of calling yourself now science and now empirical science; and, as the designation by which you are to be henceforth uniformly and exclusively known, adopt the name Science.'"

ADDITIONAL CRITICISM INVITED

I will not now request our 'scientific' expert to tell us in what measure, if any, the foregoing reply to his stated criticisms may have moderated his estimate of their force but I will ask, instead, whether he desires to offer additional criticisms. If I may judge from some conversations I have recently had with similar experts concerning the questions at issue, he may, I believe, be fairly represented as saying:

"I have some further criticisms to suggest but I prefer to suggest them in connection with certain questions I should like to ask you because I am not sure what your own view may be as to some of the possible implications and bearings of your proposal."

So it will be convenient to continue the discussion, as follows, in the question-answer form.

Question: "You have spoken much of the respective natures of two great propositional enterprises — the hypothetical and the categorical. There is a third one — which might be described as *the* propositional enterprise because its aim simply is to establish propositions (regardless of their kind). Is it not so? And is not this enterprise more generic than either of the others?"

Answer: "It evidently is so; and it is obvious that the third enterprise is more generic than either of the other two for it embraces them both."

A MORE COMPREHENSIVE ENTERPRISE REQUIRES A NAME

Question: "Do you agree with me that this more generic enterprise ought to be designated by some suitable name of its own? And, if so, seeing that you reject my proposal to call it Science, what name would you propose?

PANTHETICS

Answer: "I do agree that it ought to have an appropriate name. The best name for it that I have been able to invent, and I think it a fairly good one, is the term *Panthetics* (from *pan* and *thesis*). According to the interpreter's convenience Panthetics would be either the name of the two-term *class* of enterprises — Mathematics and Science — or else the name for all such activity as deliberately concerns itself with the establishment of propositions. Of such a man as Newton or Helmholz or Poincaré, eminent both in mathematics and in science, one would say: here is a man of *panthetic* genius — his intellectual splendor is the shining of a double star."

CONTENTION THAT MATHEMATICS IS A MODEL FOR SCIENCE EXAMINED

Question: "According to your definition of science, mathematics is not a scientific branch. It is conceivable, however, or at least supposable, that mathematics, even though it be not a branch of science, may yet be a model for it. What have you to say of the view that mathematics is such a model, an ideal pattern which scientific branches may legitimately aspire to copy and to which, as they approach maturity, they may more and more nearly conform?"

Answer: "I am aware that that view has been universally held for a long time, so long that 'the memory of man runneth not to the contrary.' I held it myself for many years but do so no longer. It is now perfectly clear to me that mathematics is no such model, or pattern. To see that it is not, it is sufficient to look the facts squarely in the face. For by what subtle process of refinement or legerdemain can you lessen the distinction between the essential nature of a categorical proposition and that of a hypothetical one? What is the trick by which one may hope to make the processes of observation and experiment approximate, in point of essential kind, the processes of deduction? How can the sense in which the proposition — if Fido has three ears and nine tails, then Fido has more tails than ears — is true be a pattern for the sense in which the proposition — Fido has a nose — is true? Merely to ask such questions is enough."

Question: "In spite of what you have just now said, it seems to me, an experimental physicist and not a mathematician, that there is *one* sense in which mathematics *is* a model for science (in your sense of the last term). I refer to mathematical certainty — to the trustworthiness of mathematical propositions. My opinion is supported by the words of one who was a very great mathematician, one of the greatest of all time, and a keen philosopher as well. For, speaking of the methodological refinement of modern mathematics, Henri Poincaré said nearly thirty years ago: 'One may say today that absolute rigor has been attained.' It would seem to follow that mathematics must own some propositions whose certainty is absolute. But in what you call science there is no proposition of which that can be said. Is it not, then, true that, in respect of certainty, mathematics is, at its best, a model for science?"

Answer: "Nothing in human speech is stronger than the word absolute. That is doubtless why people love to use it. What Poincaré said the other day had been often said before him in the course of more than two thousand years but it has never been verified. Euclid's *Elements* was long thought to be absolutely rigorous but was at length found to be far from that estate. In the rapid growth of Analysis following upon the inventions of analytic geometry and the calculus there were produced hosts of propositions that were regarded as absolutely certain until it was shown by Cauchy not only that many of them were not certain but that some of them were certainly false. Were Cauchy's methods worthy of absolute trust? So they were thought to be until Weierstrass came and showed them to be defective. Then 'Weierstrassian rigor' became a synonym for logical perfection and led to the words you have quoted from Poincaré. Just now, however, Weierstrassian rigor is being assailed, vigorously and confidently, by two men of first-rate ability — L. E. J. Brouwer and Hermann Weyl. It is safe to say that, both for mathematics and for science, absolute rigor and absolute certainty are genuine ideals but, as I have already said, genuine ideals, though admitting of endless approximations, cannot be attained."

The Realm of Science

Question: "Granting that both mathematics and science must be content with making endless approximations to the ideal of absolute certainty, is it not true that, of the two, mathematics runs the swifter race towards the unattainable goal of absolute certainty and so is always nearer that ideal than science is? May we not, then, maintain that, in the respect indicated, mathematics is indeed a model for science?"

Answer: "It can be readily shown that, contrary to the accepted view, both of your questions must be answered in the negative. Consider, for example, the propositions: men are mortal, $2+5=5+2$, New York is farther from the Moon than from the Rocky Mountains. Being established categoricals, such propositions, of which there are countless thousands, belong to science. Yet in respect of certainty, on the score of trustworthiness, they are not surpassed by any mathematical proposition in Euclid's *Elements* or elsewhere."

Question: "Either I am confused or you have made a slip. For you have just now cited $2+5=5+2$ as an instance of a scientific proposition. If it be scientific, then, according to your contention, it is non-mathematical. But surely that proposition is mathematical, is it not? Certainly the world thinks it is and so do I. Is not your citation of it a slip?"

PRACTICAL ARITHMETIC AND GEOMETRY ARE SCIENTIFIC, NOT MATHEMATICAL

Answer: "No slip but quite deliberate. In fact I made the citation just in the hope that you would challenge it, as you have done, and so give me an opportunity to drive home the truth in the matter. In regarding the proposition in question as mathematical you and your world are mistaken. It is to correct such age-old and world-wide errors, or the false conceptions from which they spring, that I am engaged in this discussion. The cited proposition is, as you know, merely one among a literally endless number of propositions that together make up Practical Arithmetic. But practical arithmetic — the arithmetic of the grocer, the banker, the farmer, the bushman of Australia or the pygmy of New Guinea (in so far as the bushman or pygmy has an arithmetic) — though it is a genuine branch of science, is not a branch of mathematics at all.

It is scientific because its propositions are, like the cited one, categorical and established. Nothing can be more evident than that. How established and when? You cannot reflect a little without seeing that they were established, not by deduction, not mathematically, but empirically, by observation and experiment; and were thus discovered and established long ages before mathematicians contrived to deduce them, only the other day, from a set of postulates, or premises. It is not mathematics but is the common experience of mankind that has made practical arithmetic part and parcel of the familiar homely wisdom of the world. Mathematics does not assert that $2+5=5+2$ or any other proposition of common arithmetic. What mathematics does assert (in this connection) is the recently discovered fact that such propositions q are deducible from a system of postulates p. When a q or a p is asserted, it is asserted just as the proposition that children are born of women is asserted, on the basis of experience, observation, experiment.

"And what I have been saying of practical

arithmetic is true of practical geometry also — the geometry of the mason, the carpenter, the cabinet-maker. Any such artisan, no matter how innocent of mathematics, will tell you, confidently, and in his own way, that, for example, the radius of a circle is equal to the side of a regular inscribed hexagon. The proposition he thus asserts is genuinely scientific for it is categorical and established — established empirically, by trial and observation — but it is not mathematical; it is no more mathematical than the proposition that unused muscles soften and shrink or that mules have longer ears than horses. Mathematics does not assert the proposition in question. What it does assert is that that proposition and countless kindred ones are implied by a certain set of postulates."

Question: "What you have said about the propositions of practical arithmetic and practical geometry is new to me but I see immediately and clearly that it is true. It evidently follows, and I have to own, that in our work we men of science continually employ as instruments many propositions which we have believed, quite erroneously, to be mathematical but which are, in fact, quite as strictly scientific (in the sense of experimental) as are any of the propositions established by help of them.

"Nevertheless you admit that in the course of their researches scientific men do often employ mathematical propositions, which they themselves establish for the purpose or else borrow for the purpose from the literature of mathematics. Or rather, to do you justice, I should say, not that you 'admit' the fact, but that you have stoutly and rightly insisted upon it as a fact of the greatest importance. You contend, however, that mathematical propositions thus brought into service by a physicist, for example, in his subject or by a chemist in his or a biologist in his or a moralist in his are neither physics nor chemistry nor biology nor ethics propositions. The contention is quite intelligible and the more I reflect upon it the more I am inclined to think it just. But in its bearing, or what you think its bearing, there is one point in regard to which I am in doubt. The question is: What sort of propositions may a scientific treatise properly contain? Let me be more specific. Suppose that I, a physicist, make a series of successful physical investigations in the course of which I employ certain mathematical propositions, and that I desire to publish a work setting forth my results. The question is whether the work must be restricted to such categorical propositions as state my physical findings or whether it may properly include also such mathematical propositions as were employed as means in the research."

ESSENTIAL TO DISTINGUISH BETWEEN USING AND ASSERTING PROPOSITIONS

Answer: "It is obvious, I think, that the answer depends. If it be the sole aim of your work to set forth your physical findings, it must rigorously confine itself to the categoricals stating them. But if the aim be, as it commonly and commendably is in such cases, not only to tell what the findings are but also to tell how they were obtained, then it is evident that the report not only may contain, but must contain, the mathematical propositions; for, by hypothesis, these are a part of the 'how' —

a part of the means by which your physics propositions, the categoricals, were established. In all such cases, however, your discourse should avoid even the appearance of *asserting* the mathematical propositions, as if you deemed them a portion of physical science; else you will serve to confirm and perpetuate a hoary, fundamental, well-nigh universal misconception in the theory of knowledge. In further response to your question one may say summarily:

"Just as a mathematical work may contain, but cannot *assert*, categorical propositions, no matter how well established in science; so a scientific work may contain, but cannot *assert*, hypothetical propositions, no matter how well established in mathematics."

THE PLACE OF PHILOSOPHY

Question: "You have not yet said anything of Philosophy — at all events not anything explicitly. I desire to ask an important question regarding it, a question of perhaps greater importance than any we have hitherto considered, for the answer may be decisive, in the judgment of many, regarding the acceptability or non-acceptability of your proposed definition of science. The question is this: If the proposal were adopted, would philosophy then belong to the domain of science or to that of mathematics or partly to the one and partly to the other or would it be something quite outside of both?"

Response: "The question seems to be important and it must be answered. But both the measure of its importance and just what the answer ought to be depend, obviously, on the meaning you are here attaching to the term philosophy. Before trying to answer your question I must, therefore, request you to indicate that meaning so that I may know in advance what the question really is."

Question: "I fear I cannot tell you what philosophy is, for who can? Nor can I say just what I mean by the term. I know, however, that there have always been men who have called themselves philosophers and have been so called by others. I know that they have been notorious for their endless discussions and that they have produced an immense body of literature called the literature of

The Realm of Science

philosophy. But I am not myself a philosopher nor am I ashamed of the fact. I am, as you know, a physicist. Like most other physicists and like most other 'scientific' men I have had little or no interest in philosophy and am as little versed in its literature as are most philosophers in the literature of what I and my colleagues are wont to call science. So I do not pretend to be able to define the term philosophy. Moreover, I do not feel obliged to attempt it, for in this discussion the definer's rôle is yours. It is you who is submitting definitions. If the term philosophy *has* a definable meaning and if you will be good enough to state it, I will adopt that meaning in the question I have asked you regarding the relation of philosophy to mathematics and science as you have defined these two terms, and I will then be glad to hear your answer to my question as thus construed by yourself. That is, I think, a fair proposal. Do you not think so, too?"

Answer: "Inasmuch as you have frankly confessed your inability to interpret your own question and have at the same time agreed to adopt such an interpretation as I may be able to give it, I think your proposal is fair enough and I will proceed accordingly.

"What, then, are we to understand your question's term philosophy to mean? An eminent French philologist was asked whether he could state in a few words just what the term philology signifies. 'That,' he replied, 'is easy; the term signifies what I am doing.' Taking that *mot* as a clew I am going to say, for the purpose of this discussion, that the term philosophy signifies *that which philosophers are doing*. I like the formula, not merely because it is true, but especially because it is not based upon the false assumption that the term's meaning is and has been everywhere and everywhen the same. For the fact is that the meaning of the term is a function of two variables — time and clime. The kind of activity in virtue of which certain men of a given time and place are then and there called philosophers is precisely what the term philosophy then and there signifies. What the term signified in medieval Europe, for example, is the distinctive type of thinking that was done by those men in medieval Europe who were then and there recognized as philosophers. What the term signifies in India today is the kind of thinking characteristic of those Indian thinkers who in their country are today ranked as philosophers. And so on. The meaning of the term obviously depends on time and place and may vary with either or both, the reason being that philosophy is, as said, the distinctive activity of philosophers and that the character of such activity is itself a function both of time and of place. One who would know what the activities were that constituted philosophy in the past or what the activities are that constitute it today must do what you say you and most of your 'scientific' brethren have not done — he must, that is, acquire at least some fair acquaintance with the history and the literature of philosophic thought.

"With these considerations in mind we may readily answer your question. Were we to construe the question with reference to the activity which the term philosophy signified during the course of many by-gone centuries, the answer would have to be very different from what it must be if we take the term in its present signification. Let me explain. There was a time, a very long period of time, when the term philosophy was employed to denote the whole intellectual activity of any cosmic-minded sage, like Plato, for example, or Epicurus or Aristotle or Aquinas or Descartes or Spinoza or Newton or Leibniz. The activity of such gigantic thinkers (or that of most of them) embraced both of the basic enterprises of which we have spoken so much: they aimed, that is, at the establishment both of hypothetical propositions and of categoricals relating to any and all of the then recognized subject-matters of the world. It is evident that, were we to take the term philosophy in that historical sense, which was the grand sense of the term, the answer to your question would be this: If my proposed definition of science were adopted, philosophy would then belong partly to the domain of mathematics and partly to that of science.

"Today, however, the term philosophy does not mean what it meant in the centuries when a man of very great gifts might aspire to the dignity and gain the reputation of a universal sage. Its significance is far less. 'Leibniz,' said De Quincey, 'was the last of the universals,' and De Quincey was right. The establishment of hypothetical propositions is today no part of a philosopher's vocation. As a field of research the realm of hypothetical propositions is the field of mathematics, not in any part that of present-day philosophy. The propositions that present-day philosophy aims to establish are categorical. The field of philosophy is, therefore, contained in the immense realm of categorical propositions but it is far from being the whole of it for that realm contains many fields that are not fields of present-day philosophy. I mean such fields as those of physics, chemistry, astronomy, biology, geology, and so on. But, though the field of present-day philosophy is thus a *relatively* small division of the realm of categorical propositions, it is, in itself, a very large and diversified field. For the categoricals which it is today the vocation of philosophy to endeavor to establish are categoricals relating to such subject-matters as those of metaphysics, epistemology, aesthetics, ethics, theology, religion, jurisprudence, politics, sociology, education, and so on. The essential point is that the propositions of present-day philosophy are categorical, like the propositions of your own subject, physics. And so it is evident that, taking the term philosophy in its present-day signification, the answer to your question must be this: If my proposed definition of science were adopted, philosophy would then belong to the domain of science — it would indeed be strictly a branch, or a group of branches, of science."

Question: "What you have said respecting the meaning of the term philosophy is to me very helpful. I especially like the formula that what the term signifies in a given time and place is just the distinctive activity of such thinkers of that time and place as are then and there reputed to be philosophers. Doubtless the formula might be advantageously applied, *mutatis mutandis*, to many another outstanding term, say physics or psychology or economics or jurisprudence or ethics or theology or engineering. In keeping with my promise

I gladly adopt your interpretation of my question and I own that the answer you have just now given to it is the answer I should have expected. But in view of that answer I am puzzled in trying to see how you can have any sufficient ground for believing that your proposed definition of science can win general approval among 'scientific' men. Are you not aware of their habit and temper? Do you not know that the attitude of most of them towards philosophy is that of indifference or that of contempt? Do but think what you are doing. You are proposing to these men such a definition of science as would require them to regard philosophy quite seriously as a genuine branch of science. Unless I am much mistaken the proposal will strike them as absurd, ridiculous, almost insolent. For the effect of such a definition, they will say, would be either an ignominious degradation of science or else a preposterous elevation of philosophy above the level of its merits."

Response: "I have long been well aware of your average 'scientific' man's poor opinion of philosophy for he has seldom missed an opportunity to express it either in the way of manifest indifference or in the haughtier form of avowed contempt. The late William James stated a well-known fact when in his last book he said: 'Down with philosophy! is the cry of innumerable scientific minds.' It has not seemed to me, however, that the contempt in question can be justified by any evident superiorities of 'science' or any evident inferiorities of philosophy; on the contrary it has seemed to me to betoken the provincial-mindedness of a too narrow specialization; and I do not believe it can survive the advent in 'scientific' circles of a truly liberal and magnanimous education. If I am here in error, I shall be very glad if you, as physicist and spokesman for 'scientific' men in general, will set me right."

Question: "Personally I am not yet certain whether your judgment in this matter is right or wrong. It is not difficult, however, to indicate the main considerations underlying the average 'scientific' man's poor opinion of philosophy. The main considerations are four:

"(1) Admitting that, as you have said, philosophy's propositions are, like those of physics or any other 'science,' categorical, he will say that, unlike 'science,' philosophy does not *establish* propositions but merely *asserts* them.

"(2) Philosophy, he will say, depends entirely too much upon the formal processes of ratiocination and is too little concerned with determining concrete ascertainable facts.

"(3) Because philosophy does not define the major terms of her discourse with adequate precision her utterances are, for the most part, either quite meaningless or unintelligible because of their ambiguity.

"(4) What he and I call science has a definite method — the laboratory method of observation, experiment, hypothesis and verification — but philosophy has not.

"Such are, I believe, the major counts in the general indictment. They are far from trivial. What have you to say of them?"

Response: "I will take them up in the given order and will try to be brief.

"(1) What is meant by an established proposition? I have already answered the question and will repeat the answer: An established proposition is one that is so regarded, so spoken of, so treated by all or nearly all expert authorities in the subject or the field to which the proposition belongs. That is what the term means in mathematics, in what you are accustomed to call science and in what I have proposed to call science. The term being thus understood, no one having even a slight acquaintance with the history of philosophic thought can maintain for a moment that philosophy has never established propositions. Consider, for example, the Aristotelian doctrine of the syllogism. That doctrine, as you doubtless know, is a product of philosophy. It involves many propositions, all of which have been established propositions for over two thousand years. That instance is alone sufficient to disprove the first count in your indictment. Yet it is only an outstanding instance among a host of like examples. Consider such propositions as the following taken from here, there, and yonder, almost at random:

"A community of humans cannot exist without some kind of government.

"Any form of (human) government involves in one way or another at least three functions, the legislative, the judicial and the executive.

"The world was created and is sustained by an all-wise and all-powerful God.

"Human beings have souls destined to survive the death of their bodies.

"It is a duty of children to obey their parents.

"It is wrong to commit murder.

"The rights of a state are superior to those of the individuals dwelling in it.

"Many miracles were performed in the course of the ages.

"Humans have minds enabling them to perceive, remember, forget, imagine, conceive, feel, reason and will.

"Man is born totally depraved.

"Faith is essential to salvation but reason is not.

"Space is an infinite container unaltered by any changes occurring within it.

"Self-preservation is the first law of nature.

"Entia non sunt multiplicanda praeter necessitatem.

"Ni la contradiction n'est marque de fausseté, ni l'incontradiction n'est marque de vérité.

"Great treasure halls hath Zeus in Heaven
From whence to man strange dooms be given,
　　Past hope or fear.

"The propositions of the foregoing list are only a few fairly random samples from a practically inexhaustible miscellany of propositions of which we have to say that they are not scientific in your sense of the term but are philosophic; that all of them were, for long periods of time, established propositions; and that many of them are still in that estate. To say that philosophy, though it asserts, does not establish, propositions is — I was about to say silly but will content myself with saying *obviously untrue*.

"(2) Being a physicist you know that the advancement of what you call science depends both upon observation (including experimentation) and upon reasoning, or ratiocination, and depends upon them equally because essentially; it depends, that is, in equal measure upon what Thomas Huxley happily called the collocation and the colligation of facts. In the old days when all serious thinking was called philosophy, the importance of reason, because it was so evident, was discovered before that of observation, and so, quite naturally, numerous philosophers, especially in those days, did, as you have said, and some still do, overstress the importance of ratiocination. It is safe to say, however, that, in respect of their number, philosophers of that stripe have been and are fully matched in 'scientific' circles by men who, depending too exclusively upon observation, have been little more than mere fact-gatherers, not interpreters of fact.

"(3) Your representative man of 'science' complains because philosophy does not define the major terms of its discourse with adequate precision. In relation thereto I have to say three things.

"First. As I have already pointed out, all discourse, whether that of mathematics or that of philosophy or that which you call scientific, ultimately is and must be discourse about matters which, though endlessly describable, do not admit of precise definition.

"Second. The matters with which philosophy deals are, for the most part, more difficult to describe well and more difficult to define approximately than are the matters with which what you call science deals. But thought about the former matters has not, on that account, less importance or less dignity than thought about the latter.

"Third. People who live in glass houses should not throw stones. Take biology, for example. The other day in conversation with a distinguished botanist and a distinguished zoologist, I requested them to tell me just what they mean by their terms *plant* and *animal*. They informed me that the terms have never been defined precisely. It so happened that these biologists were then engaged in studying the same variety of organisms. The botanist, however, regarded them as plants while the zoologist regarded them as animals. Biologists discourse extensively about continuity of organic development, about environment, and about heredity. Where in the literature of biology may one find a precise definition of continuity or of environment or of heredity? Indeterminate terms of great importance abound in that literature.

"The like is true of every other branch of what you call science. Take your own subject, physics. For generations no term figured more prominently in the literature of physics than the term ether; yet no one ever succeeded in defining it satisfactorily; and the same is to be said of many other important terms — matter, time, space, measure, magnitude, position, molecule, atom, electron, field, and the like. A very impressive revelation of such indeterminates is found, as you doubtless know, in Professor Bridgman's *The Logic of Modern Physics*. Precision of definition is, of course, of prime importance in the discourse of reason, but the lack of it is hardly more conspicuous or more baffling in the literature of philosophy than in the literature of what you are wont to call science.

"(4) Finally, your objector objects to calling a philosopher a man of science on the alleged ground that the philosopher's work is not experimental or laboratory work. It must be granted that, in the main, it is not laboratory work in the objector's sense of this term. It is my belief, however, that his sense of the term is very much too narrow. Rightly conceived, a laboratory is not merely a house or a room equipped with many ingenious technical instruments. Wherever a man, woman or child *thinks* — wherever a human being observes, identifies, remembers, imagines, conceives, discriminates, compares, analyzes, combines, reasons and judges — there is a laboratory; and it is my contention that, when the thinking aims, consciously or unconsciously, to establish some categorical proposition respecting no matter what subject-matter or aspect of the world, the laboratory ought to be regarded a scientific one. In point of aim W. B. Smith's *Ecce Deus* or J. T. Shotwell's *History of History* or W. N. Polakov's *Mastering Power Production* or V. G. Simkhovitch's *Towards the Understanding of Jesus* or Korzybski's *Manhood of Humanity* or Haywood and Craig's *History of Masonry* is as genuinely scientific as Newton's *Principia* or Weyl's *Space, Time, Matter*.

"Herewith, unless you have additional questions or criticisms, I close my defense of the proposed definition of Science. You objected to it on two grounds: on the ground that it was too narrow because it excludes mathematics from the domain of science; and on the ground that it is too broad because it includes philosophy. It is for you and your colleagues to judge whether the objections have been successfully met."

Question: "I have no further criticism to offer at present. You expressed the desire that no one would either reject or adopt the proposal before considering it maturely with open mind. Nothing could be fairer. Yet the condition is not an easy one. The proposal's first effect upon me was shocking and so I found it impossible to *begin* its consideration with a quite open mind. That effect, however, quickly subsided and thereafter, throughout our long discussion, I have both spoken and listened with fair dispassionateness, certainly without conscious prejudice. I have in candor to own, and I have pleasure in owning, that the merits of the proposal now seem to me much greater, and the objections to it much less serious, than I had at first imagined. However, I feel that I have not yet considered the matter quite maturely and so must own that I am still in doubt. I trust that you will not be disappointed if I reserve the question, as I desire to do, for further consideration."

Response: "On the contrary I am very much gratified. To expect a better outcome at the present stage would be unreasonable."

INDEX

INDEX

Absolutes as mathematical ideals, 102
Actuality, world of, x, 8, 102
Aim, of mathematics, 135; of science, 135; the mathematician's, 161-163; the scientist's 161-163
Algebra, mathematics fused with subject-matter, 54; *see* Geometry
Alibi, defense by, involves mathematical thinking, 42
Answers are propositions, ix
Applicability of mathematics exemplified in biology, 62; in ethics, 64; in theology, 65; in mysticism, 65; in law, 66
'Applied' mathematics, a misleading term, 67; *see* 'Pure' mathematics
Aquinas, 186
Aristotle, ix, 8, 186, 192
Arithmetic, part of science, not of mathematics, 175-177; *see* Geometry
Assert: To use a proposition is one thing, to assert it is another, 180-181
Autonomous whole, system, or doctrine, 7

Background of the discussion, amplitude of, fundamentality of, 20
Behavior, mathematics a cardinal type of, 81
Biblical account of creation, 43
Biology, a mathematical proposition in, 62
Bridgman, 114, 197
Brouwer, 174

Carthago delenda, general disbelief in the universal availability of mathematics, 84-91
Categorical proposition, definition of, 36; how established, 133-134; may be contained but not asserted in a mathematical treatise, 180-181
Cauchy, 174
Certainty, in mathematics, 173; in science, 175; never absolute, 174
Common sense, appeal to, 165
Content, propositional, 54, 55, 56, 57, 59, 60; may lead inference astray, 61, 62; having, and seeming to have, 70-71; *see* Form
Craig, 198
Curiosity, about the actual

world, about the world of possibility, ix, x, 7, 8

Dates, essential to the notion of established proposition, 141
Deduction, 26-28, 33, 39, 51, 58, 61, 89, 91, 155, 156, 159, 160, 161; *see* Empirical
Defined terms, their rôle, 13, 73, 93; *see* Undefined
Definition, not description, 4, 92-93; of categorical proposition, 36; of hypothetical proposition, 36; of established proposition, 141, 191; of mathematical proposition, 39; of scientific proposition, 135; of mathematics, 135; of science, 15, 135-136; of scientific method, 139; of philosophy, 184
Demonstration (of hypotheticals), 39, 58
De Quincey, 187
Descartes, 186
Description, not definition, 93; never complete, 94; as a means to indicate subject-matter, 96; explicit and implicit, 96-98; of science, 4, 128
Dialectic proof of general availability of mathematical thinking, 86-91
Discourse, ultimately about undefined terms, 101-102

Doctrinal function, 83; doctrine as an interpretation of, 83
Domain of science, mathematics excluded, 146-148, 150-151, 154, 156, 158-167, 168-169; philosophy included, 188-198
D-sense, of true, of false, 32-35, 37

Edifice, of hypotheticals, of categoricals, 163
Einstein, 149
Ellis, Havelock, 127
Emotive function of speech, 106-108; *see* Logical function
Empirical, 34, 50, 134, 155, 157, 159, 164, 166, 168, 172, 177, 178, 197, 198; *see* Deduction
Epicurus, 186
Equivalent definitions of categorical, of hypothetical, 37
E-sense, of true, of false, 32-35, 37
Established propositions, time involved, 141
Ethics, mathematical propositions in, 64
Eternal, verities, 99; implier, 161
Euclid, 12, 47-49, 77, 96-98; his 'definitions' of elements are merely partial descriptions, 97

Fact-gatherers, 19, 195; *see* Ratiocination

False, the D-sense of, the E-sense of, 33-37; *see* True

Fate, a right sanctioned by, 87, 89

Form, propositional, 54-56; deduction depends exclusively on, 57-60; *see* Content

Formulations of the idea of science, 126-128; the literature characterized, 128-130

Functions, propositional, 55; doctrinal, 83; major of speech, 106-108

'Fundamentalist' may feel the impulse to think mathematically, 43-44

Fusion of mathematical thinking with subject-matter, 54-62, 100

General applicability of mathematical thinking shown dialectically, 86-91

Geometry, the carpenter's, is part of science, not of mathematics, 177-178; *see* Arithmetic

God, a meaningless statement about, 26

Having, and seeming to have, content, 70-71

Haywood, H. L., 198

Henry, Joseph, 128

Hilbert, 77-81, 83, 98

Human, mathematics is, not divine, 100; search for the idiosyncrasy of mathematical thinking humane, 59

Huxley, Thomas, 127, 194

Hypothetical proposition, definition of, 36; how established, 133-134; may be contained but not asserted in a scientific treatise, 180-181

Hypothesis, in mathematics, 56; in science, 155-156, 161; *see* Implier and Postulates

Ideal, nature of, 116; service of, 116-117; of logically perfect language, 110-116

If-then propositions, some hypothetical, some categorical, 9, 36

Implication, nature of, 27, essential to mathematics, 10; a relation between propositional forms (not content), 162; essential to inference, deduction, logical organization of thought, 27

Implicate, 36

Imply, 10

Implier, 36, 161; *see* Hypothesis and Postulates

Indifinables, 101-102

Indeterminate terms, in science, in philosophy, 19, 191, 196-197

Inference (logical), depends on propositional form, not on content, *see* Deduction and Implication
Interest, the mathematician's, the scientist's, contrasted, 160-163
Interpretation of doctrinal function yields doctrine, 83
Intervention of mathematics in science, 158-164

James, William, 190
Jurisprudence, a mathematical proposition in, 66

Keyser, 26, 55, 78, 83, 102, 117, 118
Knowledge-seeking activity, its grand divisions, 9, 114-115, 138
Korzybski, 198

Laboratory, nature of scientific, 198
Lagrange, 149
Law, municipal, *see* Jurisprudence
Leibniz, 24, 186, 187
Logical function of speech, 107-115; *see* Emotive function

Magnanimity, required, 20
Mathematical proposition, definition of, 39

Mathematical symbols, essential to the growth but not to the existence of mathematics, 44-45
Mathematical thinking, dependent on no kind of subject-matter, applicable to all kinds, 41-44, 86-91
Mathematics, definition of, 24, 135
Maxwell, 149
Method of science, the, 139
Mot, Russell's famous, 76
Mysticism, a mathematical proposition in, 65
Myths, ages-old, regarding the nature of mathematics, 40-45

Natural bases for defining mathematics and science, 8-9
Newton, 141, 149, 198
Nonsensical, meaningless statements and discourse, 25-26

Ogden, 111
Open-minded, 130, 144, 199
Ordinary words, primitive implements for mathematics, 44-45
Orientation in the geography of thought, 143
Ought, special meaning of, 5
'Overarching presence in the sky,' 66

Panthetics, 171

Pasteur, 126
Pearson, 127
Philosophy, definition of, 184; function of time and place, 184-188; relations of, to mathematics and to science, 186, 188
Pieri, his conception of mathematics, of mathematical thinking, 24
Plato, 23
Poincaré, 127
Polakov, 198
Possibility, the world of, x, 8, 102
Postulates, nature and rôle of, 48, 50, 74, 75, 78, 89, 91; *see* Hypothesis and Implier
Pragmathetics, 168
Pragmatic, 50, 168
Proposition, nature of, 25-26
Propositional form, content, 54-56; function, 55; types, *see* Categorical and Hypothetical
'Pure' mathematics, misnomer, 67-70; *see* 'Applied' mathematics

Quest of categoricals, of hypotheticals, 8
Question-asker, *see* Curiosity
Questions, concerning the actual lead to categorical propositions and so to science, 8, 136; concerning the possible lead to hypothetical propositions and so to mathematics, 8, 135
Radical difference between the two meanings of true (or false), 33-37; between mathematically and scientifically established, 155; between mathematical and scientific hypothesis, 161; between the aims and interests of the mathematician and those of the scientist, 161-166; between collocation and colligation of facts, 194-195; *see* Fact-gatherers and Ratiocination
Ratiocination, 191, 194; *see* Fact-gatherers
Realm of categoricals, of hypotheticals, x, 134-135, 137-138
Richards, 111
Riemann, 127
Rôle of genuine ideals, 116-117
Ruskin, 126
Russell, 55, 76

Science, often described but never defined, 4-5, 120-122, 126-130; proposed definition of, 15, 135-136
Scientific hypothesis, nature of, not same as mathematical, 155-156, 161
Scientific method, definition of, 139

Scientific proposition, definition of, 135; how established, 155
Seeming to have, and actually having, content, 70
Senses, the two, in which a proposition may be 'about a kind of subject-matter,' 164-165
Sheer mathematical thinking deals with no kind of subject-matter, 53-61
Shotwell, 198
Smith, W. B., 198
Speech, the major functions of, 106-115
Spinoza, 186
Spurning of philosophy by 'scientific' men, a token of provincial-mindedness, 190
Stone age of mathematics, 70
Summary, partial, 98-102
Symbols, *see* Mathematical symbols

Theology, mathematical propositions in, 65
Townsend, 77
True, the two meanings of, 32-35, 37; *see* False

Truth-seeking activity, its grand divisions, 136-137

Undefined terms, 72-74; their role, 74, 86-87; how best denoted, 75; where they enter, 75-76; basis of all discourse, 101

Varieties (of subject-matter), countless, unified by form, *see* Doctrinal function

Weierstrass, 174
Weierstrassian rigor, 174
Weyl, 174
Whetham, 127
Whitehead, 127
Wisdom, origin of arithmetical, 176-177; *see* Geometry
Wonder, ix; *see* Curiosity
World, without implication would be subhuman, 27

Years, a half million, required to ask, and two thousand to answer, a certain question, 23

COLUMBIA UNIVERSITY PRESS
Columbia University
New York

FOREIGN AGENT
OXFORD UNIVERSITY PRESS
HUMPHREY MILFORD
Amen House, London, E. C.

Bei Fragen zur Produktsicherheit wenden Sie sich bitte an:
If you have any questions regarding product safety,
please contact:

Walter de Gruyter GmbH
Genthiner Straße 13
10785 Berlin
productsafety@degruyterbrill.com